Large-Scale Energy Models

Prospects and Potential

AAAS Selected Symposia Series

Published by Westview Press, Inc.
5500 Central Avenue, Boulder, Colorado

for the

American Association for the Advancement of Science
1776 Massachusetts Ave., N.W., Washington, D.C.

Large-Scale Energy Models

Prospects and Potential

Edited by Robert M. Thrall, Russell G. Thompson, and Milton L. Holloway

AAAS Selected Symposium **73**

AAAS Selected Symposia Series

This book is based on a symposium which was held at the 1980 AAAS National Annual Meeting in San Francisco, California, January 3-8. The symposium was sponsored by the Operations Research Society of America, the Society for Industrial and Applied Mathematics, the Institute of Management Sciences, and by AAAS Sections A (Mathematics), M (Engineering), and P (Industrial Science).

Published in 1983 in the United States of America by
Westview Press, Inc.
5500 Central Avenue
Boulder, Colorado 80301
Frederick A. Praeger, President and Publisher

Library of Congress Cataloging in Publication Data
Main entry under title:
Large-scale energy models, prospects and potential.
(AAAS selected symposium ; 73)
"Based on a symposium which was held at the 1980 AAAS National Annual Meeting in San Francisco, California, January 3-8. The symposium was sponsored by the Operations Research Society of America ... (et al.)" --T.p. verso.
Includes index.
1. Energy industries--Mathematical models--Congresses. 2. Energy consumption--Forecasting--Mathematical models--Congresses. I. Thrall, Robert McDowell, 1914- . II. Thompson, Russell G. III. Holloway, Milton L. IV. American Association for the Advancement of Science. V. Operations Research Society of America. VI. Series.
HD9502.A2L37 1983 333.79'0724 82-20119
ISBN 0-86531-408-X

Printed and bound in the United States of America

About the Book

Large-scale energy models are becoming increasingly important as the availability of resources grows more variable and major policy changes are required more frequently. Historical trends are no longer adequate predictors of the impacts of major changes on various sectors of the economy because they do not account for possible alterations in underlying structural relationships. To predict the outcomes of such changes correctly, econometric models must be used; however, current models need significant improvements if they are to detect the full range of effects a particular policy change will have.

This book describes state-of-the-art modeling in industry and agriculture and links macro models of the economy with micro models of industrial and agricultural sectors. The authors are expert in the construction, modification, transfer, use, and evaluation of large-scale energy models.

About the Series

The *AAAS Selected Symposia Series* was begun in 1977 to provide a means for more permanently recording and more widely disseminating some of the valuable material which is discussed at the AAAS Annual National Meetings. The volumes in this *Series* are based on symposia held at the Meetings which address topics of current and continuing significance, both within and among the sciences, and in the areas in which science and technology impact on public policy. The *Series* format is designed to provide for rapid dissemination of information, so the papers are not typeset but are reproduced directly from the camera-copy submitted by the authors. The papers are organized and edited by the symposium arrangers who then become the editors of the various volumes. Most papers published in this *Series* are original contributions which have not been previously published, although in some cases additional papers from other sources have been added by an editor to provide a more comprehensive view of a particular topic. Symposia may be reports of new research or reviews of established work, particularly work of an interdisciplinary nature, since the AAAS Annual Meetings typically embrace the full range of the sciences and their societal implications.

WILLIAM D. CAREY
Executive Officer
American Association for
the Advancement of Science

Contents

About the Editors and Authors

Robert M. Thrall *is professor of administration and Noah Harding Professor of Mathematical Sciences at Rice University. A specialist in decision analysis and game theory, he has written numerous articles and edited several books in this field, including* Mathematical Frontiers of the Social and Policy Sciences *(edited with L. Cobb; AAAS Selected Symposium 54, Westview, 1981),* Decision Information *(edited with C. P. Tsokos; Academic, 1979), and* Economic Modeling for Water Policy Evaluation *(Elsevier, 1976).*

Russell G. Thompson *is professor of quantitative management sciences at the University of Houston and president of Operational Economics, Inc., a data analysis and forecasting firm in Houston, Texas. He has been the technical manager of many complex interdisciplinary projects to develop sound methods of evaluating the structural economic consequences of changes in interdependent government policies. His research areas are water, energy, and environmental and regional economics. He has published many articles and has been instrumental in the publication of four computer-based analysis books:* The Cost of Clean Water, The Cost of Electricity, The Cost of Energy and a Clean Environment, *and* An Economic Model for U.S. Oil and Gas Supplies.

Milton L. Holloway, *a resource economist, is executive director of the Texas Energy and Natural Resources Advisory Council in Austin. His publications include* Texas National Energy Modeling Project: An Experience in Large-Scale Model Transfer and Evaluation *(Academic, 1980),* A Study of the Texas Energy Advisory Council: Its Structure and Functions Relative to State Science, Engineering and Technology Transfer *(with R. J. King and M. Stevens; 1979), and* Texas Energy Outlook: The Next Quarter Century *(Austin: Governor's Energy Advisory Council, 1977).*

C. West Churchman *is professor emeritus of the School of Business Administration at the University of California, Berkeley, and adjunct professor of the Associated Faculty in the Social Systems Sciences Department, the Wharton School of the University of Pennsylvania. Originally trained in symbolic logic and philosophy, he has ranged widely, from applied statistical methods, city and health planning, education and R&D management to accounting, marketing, economics, and social and moral issues in business and government. He organized the first academic operations research group, and in 1957 wrote the first textbook,* Introduction to Operations Research. *Among his recent works are* The Systems Approach *(Dell, 1968; rev. ed., 1979),* The Systems Approach and Its Enemies *and* The Design of Inquiring Systems *(Basic, 1979 and 1971, respectively).*

George B. Dantzig *is professor of operations research and computer science at Stanford University. A specialist in mathematical programming and applicational mathematics, his publications include* Linear Programming and Extensions *(Princeton, 1963) and* PILOT-1980 Energy Economic Model, Vol. 1: Model Description *(with B. Avi-Itzhak and T. J. Connolly; EPRI, 1980). His work has earned him numerous awards, among them the National Academy of Sciences Award in Applied Mathematics and and Numerical Analysis, the National Medal of Science, and the John von Neumann Theory Prize.*

C. Roger Glassey *is professor and chairman of the Department of Industrial Engineering and Operations Research at the University of California, Berkeley. Formerly assistant administrator for Applied Analysis, Energy Information Administration, Department of Energy, he has conducted an analysis of U.S. energy policy using large-scale models.*

A. N. Halter *is an energy economist on the technical staff of the Department of Energy Analysis at the Electric Power Research Institute in Palo Alto, California. He is the author of* Decisions Under Uncertainty with Research Applications *(South-Western Publishing Co., 1971).*

Earl O. Heady *is director of the Center for Agricultural and Rural Development, Charles F. Curtiss Distinguished Professor of Agriculture, and professor of economics at Iowa State University. He is the author of 21 books and more than 750 articles, and his research has earned him numerous fellowships and awards, most recently the American Agricultural Economics Association Award for Outstanding Research.*

William W. Hogan *is professor of political economy (the IBM chair, Technology and Society) and director of the Energy and Environmental Policy Center at John F. Kennedy School of*

Government, Harvard University. He has done extensive research on energy modeling, and his current research focuses on energy and security, decision analysis applied to synthetic fuels, and the role of analysis in the policy process. He has published widely in his field, and his books include Energy Information: Description, Diagnosis, and Design *(Stanford Institute for Energy Studies, 1979).*

Bart Holaday, *currently president of the Tenax Oil and Gas Corporation in Houston, Texas, has been involved in energy analysis for more than a decade. He has served as vice president for planning at Gulf Oil Exploration and Production Company, deputy assistant administrator for analysis at the Federal Energy Administration, and director of the Office of Energy Data and Analysis at the Department of the Interior.*

Charles C. Holt *is director of the Bureau of Business Research and professor of management at the University of Texas, Austin. An economist with training in electrical engineering, he has done research involving modeling and decision analysis.*

Wen-yuan Huang *is an agricultural economist at the U.S. Department of Agriculture in Ames, Iowa. A specialist in agricultural production and resource economics, he has done research on econometric models for agricultural policy analysis and the software tools for their use.*

Lawrence R. Klein, *currently Benjamin Franklin Professor of Economics at the University of Pennsylvania, has specialized in econometrics and macroeconomic theory and policy. His work has earned him numerous prizes and awards, including the Nobel Prize in Economics. He is the author of numerous books and articles in his field, among them* An Introduction to Econometric Forecasting and Forecasting Models *(with R. M. Young; Lexington, 1980) and* An Econometric Model of the United States, 1929-1952 *(with A. S. Goldberger; North Holland, 1955).*

Lincoln E. Moses, *professor of statistics at Stanford University, has done research on statistical methodology, medical applications of statistics, and quantitative methods in public policy. He was the first administrator of the Energy Information Administration at the U.S. Department of Energy.*

Sankar Muthukrishnan, *an operations research analyst at Texas Instruments in Houston, has done research on energy, econometrics, financial modeling, and performance and resource modeling.*

Srikant Raghavan, *a specialist in quantitative methods and operations research, is a consultant with Tata Consultancy Services in Bombay, India.*

Michael Rusin, *a specialist in energy economics, is manager of policy analysis at the American Petroleum Institute in Washington, D.C. His publications on energy-related topics have appeared in* Chemical Engineering Science, Industrial and Engineering Chemistry Fundamentals, *and the* Journal of the Society for Industrial and Applied Mathematics.

John C. Stone, *an economist with Operational Economics, Inc., in Houston, Texas, has published on linear programming models of industry and on partial equilibrium models of the energy sector.*

James L. Sweeney *is director of the Energy Modeling Forum, chairman of the Institute for Energy Studies, and professor of Engineering-Economic Systems at Stanford University. A specialist in energy modeling and analysis, and former director of the Office of Energy Systems Modeling and Forecasting, Federal Energy Administration, he has studied the economics of depletable resources, gasoline demand forecasting, and petroleum product price forecasting.*

Reuben N. Weisz *is an operations research analyst in Land Management Planning, U.S. Forest Service, Albuquerque, New Mexico. He has worked in the areas of economics, agriculture, water resources, and on the use of a recursive adaptive programming model in analyzing agricultural policies.*

Preface

This book, *Large-Scale Energy Models: Prospects and Potential*, is based on a three-session symposium in the January 1980 American Association for the Advancement of Science (AAAS) annual meeting in San Francisco. Since the speakers were selected from government, industry, and academia, it is not surprising that there were keen debates on a number of issues.

The three of us who organized the sessions had worked together on the transfer of the Federal PIES/MEFS model from Washington, D.C., to College Station, Texas (some of the details of which are described in Part III of this book), and we felt that problems and observations generated by this transfer were worth airing in a broader setting. Most of the authors were directly involved in this transfer operation, some at the State and some at the Federal level.

Sessions 1, 2, and 3 of the symposium appear, respectively, as Parts I, II, and III of this book. The five chapters in Part I present a variety of viewpoints on modeling and are not as narrowly focused as those in the other Parts. Part II has three chapters on various aspects of the need for linkages between models at different hierarchical levels. The first four chapters of Part III discuss the importance of model documentation and transfer as well as some alternatives to transfer. The final chapter reviews the general status of quantitative modeling for energy analysis. Each Part has its own introduction.

Robert M. Thrall,
Russell G. Thompson,
Milton L. Holloway

Part 1

Some Views on Large-Scale Energy Modeling

Robert M. Thrall

Introduction: Part 1

Although managerial enthusiasm for use of analytic models has not generally been high, there are enough exceptions with successful outcomes to encourage modelers to continue their development. As one result of their interactions both modelers and managers have become more sophisticated as well as more realistic in their demands and expectations.

It has been argued that an "ideal" model should be both comprehensive and simple, both derived from sound theory and based on reliable and adequate data. In view of these somewhat contradictory requirements it is not surprising that there are few "ideal" models extant. The modeler must learn to balance these factors appropriately in terms of the problem at hand.

The five chapters of Part I all relate to this balancing act. Lincoln Moses makes a strong case for simple models. George Dantzig agrees in part but also presents a case for large, complex models. Bart Holaday derives from his experiences in both government and industry a set of conditions for effective utilization of models by managers. West Churchman discusses inclusion of regional (or other local) factors in modeling. In chapter 5 Michael Rusin illustrates some of the problems of energy modeling using a case study related to the windfall profits tax.

An issue not made explicit in these chapters but which needs more attention is the role of non-monetary benefits and costs in management decisions. Churchman touches on this point in his emphasis on the importance of local factors in modeling. An emerging research area, "multi-criterion decision analysis" may have developed concepts appropriate to energy modeling. Inclusion of non-monetary factors, such as quality of life and political inability, could substantially enhance a model's value for policy decisions. The importance of quality of life considerations is illustrated by public reaction to the gasoline shortage of 1979, which was apparently based more on personal inconvenience than on price per gallon, and which generated sufficient political heat to stimulate action at both state and federal levels.

Although there is general agreement that the usefulness of a model is enhanced by its simplicity, the question remains as to how one best reaches a viable, simple model. One strategy, familiar in statistics, is to start with a few of the major variables (factors) and add new variables one at a time until the model fits the underlying data acceptably.

A second strategy is to construct an initial model that includes all of the potentially useful variables and combines them according to the best available knowledge of their actual effects. This initial model will ordinarily be too cumbersome for practical use; therefore, the modeler proceeds through a sequence of simplifications obtained by discarding certain variables, combining others, and using approximations of overly complex terms.

The first strategy may be described as "bottom up" and the second as "top down." (Note that these terms are used in a different sense in Part II.) The top-down strategy is less straightforward but may be less likely to overlook important variables or variable interactions. Another advantage of the top-down strategy is that the initial model can be profitably used as the starting point for many different situations.

Part I only scratches the surface of general considerations of large-scale modeling for energy or other policy decisions. Clearly, the dialogue will continue.

Lincoln E. Moses

1. Energy Models: Complexity, Documentation, and Simplicity

Energy policy modeling is a complex undertaking. Energy models address intricate phenomena. Energy is supplied in many ways from many different sources. The demand for energy is enormously dispersed and varied. Energy policy is understood through economic theory, which lacks the clarity and soundness of Newton's Laws. The processes occur within a social and legal context fraught with regulatory constraints and interventions. Questions addressed through energy models often are in the form "If a proposed change is made in a certain law, or tax, or tax benefit, or interstate freight rate, what will be the effect on the price of coal, or electricity, or both, in some particular region of the country?" Unless the model contains in its structure representations of both the queried intervention and the queried outcome, it must remain silent on that question. Hence, a model designed to analyze energy policy questions probably has elaborate features simply because of its intended uses. In addition to being complex, energy models are likely to have problematic data, if only because energy statistics (as well as energy modeling) have a rather brief history.

To these substantive difficulties may be added the highly controversial quality of many energy policy issues. Complexity, data problems, and inherent controversiality together impose a burden of skepticism on any (unwelcome) model results. The first attacks on undesired results are likely to be technical cavils about the model and its supporting data. At this juncture the first line of defense is the existence of clear, sound documentation of both the model and its data. The persuasive power and intellectual cogency of a model cannot exceed the quality of its documentation. In a broad sense, full documentation includes the related processes of intermodel comparison, sensitivity analysis, external review, and portability ensurance.

All of these activities are time consuming and costly, and they compete for the modeler's attention, which he or she wants to apply

Table 1

Variability of a sum containing one component with nonvanishing error

σ_1	σ_2	σ_3	σ	%
1.0	.5	.5	1.22	100
1.0	.4	.4	1.11	91
1.0	.3	.2	1.06	86.6
1.0	.1	.1	1.01	82.5
1.0	.01	.01	1.0001	81.7
1.0	0	0	1.000	81.6

to other projects. But the demands for documentation, review, comparison, and portability are legitimate and persistent.

A key to meeting these demands--and often an aid to better modeling--is to favor simplicity rather than complexity in constructing energy models. There are limits to simplification, of course, but my view is that simplification should be pursued.

The result of simplification is not only response to critics and other interested parties; in addition, there may result a reallocation of the analysts' intellectual effort. Operating a large, complex model can require very large staff effort in simply "getting it to run." There is a danger that a disproportionate share of labor and thinking will go into running the model and explaining end results. The alternative is front-end thinking, formulating the issues, conceiving various analytical approaches, deciding on alternatives, and so on. Further rewards of simpler models include (1) greater ease of sensitivity analysis of nonpolicy variables as a way of exploring forecast uncertainty, and (2) more opportunity to attack a problem from two or more analytic approaches, thus exploring uncertainty arising from model formulation.

I now turn to some technical observations about simplicity in modeling. First, it is worth recognizing how little may be gained by reducing inaccuracy in some elements if others cannot also be improved. For example, let X denote the sum of three statistically independent components, X_1, X_2, and X_3. The uncertainty in X can be expressed by its standard deviation σ, which is related to the uncertainties of the three components by the expression

$$(1) \quad \sigma^2 = \sigma_1^2 + \sigma_2^2 + \sigma_3^2 \quad \text{or} \quad \sigma = (\sigma_1^2 + \sigma_2^2 + \sigma_3^2)^{\frac{1}{2}}$$

Now suppose that $\sigma_1 = 1.0$, $\sigma_2 = .5$, and $\sigma_3 = .5$. Then as shown in Table 1, $\sigma = 1.22$. Suppose we find some means to reduce σ_2 and σ_3 (but not σ_1); the effect will be to reduce σ. However, the effect is disappointingly small, as successive lines of the table show. Even if we completely remove all uncertainty from X_2 and X_3, reducing σ_2 and σ_3 to zero, the value of σ falls only to 81.6% of its original value.[1] The example illustrates two principles: large sources of error dominate the composite, and reducing smaller error elements may do little good.

There is worse news about inessential detail: it can hurt appreciably. As a real life example, twenty-five years ago the quantitative measurement of thyroid-stimulating hormone (TSH) could be done only by very tedious, expensive means. A colleague-

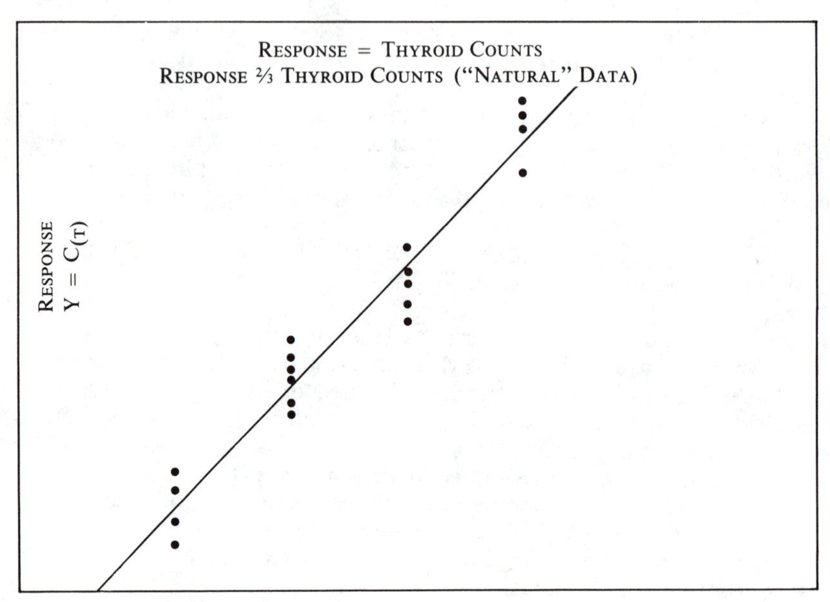

Figure 1. Average Response to Increasing Dosage.

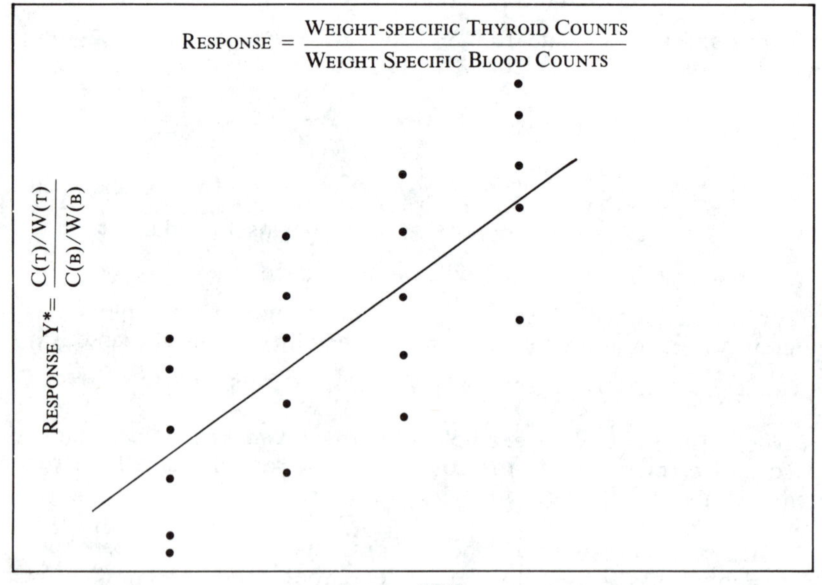

Figure 2. Average Response to Increasing Dosage.

client was trying to develop a less tedious way, a bioassay using day-old chicks. The idea was that injection of TSH would stimulate changes in the chick's thyroid that would manifest themselves by increasing the thyroid's capture of iodine from the circulating blood. The rate of iodine capture could be detected by injecting the chick with I^{131}, a radioactive form of iodine. In Figure 1, each dot represents one chick thyroid. The response, counts of the radioactive emitter observed in the thyroid, indeed grew with dosage of TSH. This led to the hope of measuring the unknown concentration of TSH in some sample of interest, by seeing how many counts it produced, and referring that outcome to the line based on known doses.

The investigator was not entirely happy with using simply the counts, C (T), for the response variable, Y, because different chicks have different sized thyroids and the activity was a qualitative difference that might get mixed up with the mere weight of the thyroid. So he thought it might be better to use a specific activity, C(T)/W(T) where W(T) is the weight of the thyroid. Since the dose of radioactive iodine was not carefully measured, and more important, it was not injected into the blood stream but into the chick's belly, the amount reaching the blood could be quite different in different chicks. He therefore suggested that the blood concentration of the I^{131} should be taken into account by dividing C(T)/W(T), the specific thyroid activity, by a measure of the blood concentration. That measure would be obtained by taking a blood sample, counting it, weighing it, and putting the count on a weight basis. So the blood's specific activity would be C(B)/W(B). Then the adjusted index was proposed:

$$Y^* = \frac{C(T)/W(T)}{C(B)/W(B)}$$

To assess the desirability of this more refined response measure I took the same chicks' data and plotted the results in terms of Y*. Figure 2 displays the results; compare it to Figure 1. The line is shallower; the scatter about the line is greater. The sensitivity of the assay scheme is sharply reduced. In Figure 1 there is a clear separation of response levels at adjacent doses. This is not true in Figure 2; using Y*, only the first and fourth dose levels exhibit clearly separated responses.

The critical parameter in a bioassay is σ/β, where σ denotes response variability at a fixed dose, and β denotes the slope. It turns out that the sample size needed to attain a given level of precision in such a bioassay is proportional to the square of σ/β. In Figure 2, β has 3/5 the value of β in Figure 1, and σ has 3 times the value of σ in Figure 1. Thus the critical parameter is five times as large for Y* as for Y. (This was the exact feature of the real data, which I no longer have.) Then the "refined" index Y* would require (approximately) 25 times as many observations to give precision equal to that obtainable by simply using Y, the "uncorrected" counts.

This example shows how the incorporation of inessential detail can be harmful. The whole concept was summarized by George Box (1) who wrote:

> There is a tendency to try to model each step that the investigator can imagine, whether there is strong evidence that that step really occurs in the system or not, whether the step affects the solution or not, and whether the data could possibly supply any information about that step or not. Even if he had a 50% chance of being right about any given step, the investigator need only introduce a few such steps into a system and the chance of error becomes overwhelming. My experience is that we must borrow William of Occam's razor and use it rather ruthlessly to remove deadwood. Usually, models are best built up from simple beginnings, elaboration being introduced only as it is shown to be necessary. . . .

It is fair to say that the enemy is verisimilitude ("appearance of being true, semblance of actuality"). Any useful model must omit features of reality, but it can be hard to answer a critic who says "How could you omit X, Y, Z, and H?!" (which enhance the semblance of actuality). Such criticism provides a strong impulse for the modeler to include them in a later, more elaborate model. But in fact, the answer should often be "Because X, Y, Z, and H are clearly much less important and uncertain than A, B, C, and S, which are already in the model and will dominate outcomes." Sometimes the answer may be "Because X, Y, Z, and H are even less important and more uncertain than P, R, S, and W, which have also been omitted from the model as inessential."

It may be asked why I have twice coupled the terms "important" and "uncertain." I suggest that we should experiment with a forecasting method which I shall later call "ignorance-oriented modeling." The processes composing this curiously named concept are three:

(1) Identify those key features that are influential and relatively certainly known. (Domestic crude oil production next year is an example.)

(2) Identify those key features that are influential and that carry substantial uncertainty (such as crude oil imports next year).

(3) Construct the forecast so that the uncertainty in the outputs is directly related to the key uncertain inputs.

This approach is very different from forecasting all components of the problem with a modeler's "best shot" and presenting the result with or without a sensitivity analysis.

I offer an example that I presented in October 1979 in testimony[2] before a Senate Subcommittee. Because the testimony was in early October, I chose to examine years extending from October through September (Table 2). The top two rows of Table 2 are actual history, "fiscal years" ending October 1979 and October 1978. It is striking how similar all features are between those two years. But the later year saw gas lines and sharp price increases; the earlier year was relatively trouble-free. This suggests that answering the question "will there be gas lines?" is not easy. The answer is "maybe."

In all three forecasts, domestic crude oil production is 8.6 mmbd, a figure likely to be correct to the accuracy shown. During what we now call the first Iran Crisis, the single dominant feature of the whole supply situation was the availability of imported crude. So three levels of this variable were chosen, the middle value 7.9 differing only slightly from the 7.84 and 7.88 of the two prior fiscal years (see column 3). Distillate supply could in principle oscillate quite a bit, but it is not likely to rise in years of tight supply and rising prices. If a real shortage of petroleum appears, distillate will be protected before gasoline, because the former warms homes and makes jobs. (The choice of distillate over gasoline was made as early as 1942 during World War II by Harold Ickes.) For these reasons the value of distillate is set in all three cases as essentially equal to that in the two prior years. Now what is uncertain is the amount of gasoline that will be available, and availability is keyed directly to import availability.

Who knows how good these forecasts are? They have advantages: they can be explained; they can be (directly) understood; they focus the uncertainty of the outcome back onto the dominant uncertainty of the next few months. It is indeed an "ignorance-oriented" set of forecasts. It is probably worth noting some factors that have been neglected: prices, GNP, changes in refinery plant, fuel switching, lead-free gasoline issues, and crude and product inventories. All of these are uncertain, but as long as they each carry less uncertainty than is carried by imports we are not likely to be losing much, if any, accuracy.

A second ignorance-oriented approach to examining probable gasoline supply in the next fiscal year could use world oil price (influential and uncertain) as the key variable. A third approach

Table 2

THREE SCENARIOS AND RECENT HISTORY CONCERNING DISTILLATE AND GASOLINE SUPPLY
OCTOBER – SEPTEMBER
(all quantities other than percentages are in millions of barrels per day)

(1) Actual	(2) Domestic Crude Production	(3) Net Imports Crude + Product	(4) Distillate Supply (Incl. Imports)	(5) Distillate/ Refinery Output	(6) Gasoline Supply (Incl. Imports)
FY78	8.65	7.71 (+ .13)*	3.31	19.9%	7.25
FY79**	8.60	7.75 (+ .13)*	3.35	19.8%	7.27
Scenario					
High FY80	8.6	8.2 (0)*	3.35	19.3%	7.65
Medium FY80	8.6	7.9 (0)*	3.35	19.6%	7.33
Low FY 80	8.6	7.6 (0)*	3.35	19.8%	7.04

*( ) = SPR Imports **Oct. 78 through Aug. 79

might use both imports and world oil price as major variables because their uncertainties will dominate the gasoline supply picture in the next fiscal year.

I restate the concept: put the important uncertain elements into close and evident juxtaposition with the model's outcomes. Use no more detail than is important (read "is known to be important," not "could be important"). Assessing these matters puts a premium on "front-end" thinking in contrast to "after the output" thinking, and may be difficult, although it was not in this case.

I conclude with a word about the unattractive name "ignorance-oriented modeling." It conveys an element of rebellion against the nomenclature of my own discipline, including such pleasant, promising, smiley terms as: efficient, robust, most powerful, unbiased, sufficient, complete. A more important aspect of the name is that it focuses attention on the unknowns, the things to learn more about.

Notes and References

1. This bad news can be carried much further. Suppose that there are infinitely many independent sources of error with standard deviations as shown:

$$\begin{aligned}
\sigma_1 &= 1 \\
\sigma_2 &= \sigma_3 = 1/2 \\
\sigma_4 &= \sigma_5 = 1/3 \\
\sigma_6 &= \sigma_7 = 1/4 \\
&\vdots \\
\sigma_{2i-2} &= \sigma_{2i-1} = 1/i \\
&\vdots
\end{aligned}$$

Then $\sigma^2 = \sum_1^\infty \sigma_i^2 = 1 + 2 \quad (\sum_2^\infty 1/i^2)$

$$= 2 \cdot (\sum_1^\infty 1/i^2) - 1$$

$$= 2 \cdot (\pi^2/6) - 1 = 2.2899,$$

and so $\sigma = (2.2899)^{1/2} = 1.513.$

Therefore, if the investigator eliminates (by infinite pains) all sources of error but the first, σ will be reduced from 1.513 to 1.000, a reduction of about 34%.

2. Statement of Lincoln E. Moses, Administrator, Energy Information Administration, before the Subcommittee on Energy Regulation of the Committee on Energy and Natural Resources, U.S. Senate, "Current Petroleum Situation," October 22, 1979.

(1) Box, G.E.P., J. Wash. Acad. Sci., 64, 58 (1974).

2. Concerns About Large-Scale Models

I take as my theme certain concerns about models voiced by Lincoln Moses in a paper appearing in revised form in this book under the title "Energy Models: Complexity, Documentation, and Simplicity." I will paraphrase a few of Moses' comments. He said that when he left the home grounds of statistics and wandered into energy modeling and forecasts, it was indeed a new world:

First there was a great increase in complexity. Energy models address intricate phenomena. The supply of energy comes in many forms from many sources; the demand for energy is also enormously varied and dispersed.

Second was the need to make forecasts. Energy forecasting is unlike statistics, in which extrapolations can usually be made within the domain of actual observation. Moses found that "energy modelers must say something about energy futures with sharply rising prices, while the bulk of their data comes from decades of gradually falling energy prices." Price elasticities observed under one set of conditions were being applied to a very different set.

Third was something disturbing--the acceptance by modelers of the causal relations as stated in economic theory.

Fourth, also disturbing, was the non-existence of a laboratory for testing the effects of a new policy. More precisely, the laboratory is society itself, and once a policy is tested it is too late.

Fifth was the lack of good data for the models and the fact that the world contains surprises that were not and apparently could not be reflected in any models--Kakahomania, Three Mile Island, gasoline lines, and gasoline prices rising by quantum leaps.

But the biggest surprise was that modelers were oblivious to these pitfalls; they marched ahead, waving their banner "We can model anything." Moses marveled at the chutzpah of modelers and likened them to fortune tellers and astrologers for their practice of forecasting phenomena for next year--or 1990--that we cannot, or do not, measure today.

Still another worry was the fact that energy models can have hundreds of equations and thousands of variables, constants, and parameters. There are bound to be errors and omissions that escape detection--so how can one trust the results and believe the forecasts? What about failure of models to take into account uncertainties?

Finally, Moses worried about the doctrine of neoclassical economists that the driving economic force is optimizing responses of demand and supply to a uniform price system: profit maximization by the firm; satisfaction maximization by the consumer. He questioned these premises. Do consumers "optimize" anything? How are investment decisions really made? Occasionally in a group decision-making situation an optimizer will be present; he is at risk of being seen as a nut. Instead of optimizing some objective function, decision making appears to be directed at ensuring a very low probability of any major difficulties.

To Moses' list of worries I would add my own greatest worry, that policymakers distrust models and continue to make decisions about complex systems without the benefit of models. In general I agree with Moses' concerns about models, particularly energy models, and with the steps he outlined to overcome them. On one point I disagree with him: the use of models as a forecasting tool.

My view is that the economy is a machine for producing goods and services. It has a flexible technology at its disposal. Over time the economy can evolve in many ways; in short, there are many possible futures. The modeler's job is to calculate and then to present to policymakers in broad terms the effects of a variety of possible policy decisions. Based on this menu of alternatives the policymakers' job is to come to a consensus as to which policy appears to them to be the best.

I too feel that energy modelers who use models to forecast phenomena occurring next year or in 1990 are audacious, bold, and deserve to be classed with fortune tellers and astrologers. When we decide, however, that the economy is not an enigmatic black box whose mysterious, capricious behavior we are trying to predict but rather is a vast machine whose technology is for the most part known, then we can ask the question, "how can that economy best serve us?" What broad investment, conservation, and production policies should we engage in now to get from here to there? The "there," the "where to get to," is itself a policy decision.

I will next discuss the complexity of our modern economy, the dynamic forces shaping it, and how models and computers can be used to come to grips with the economy. Harrison Brown, in his book, Learning How to Live in a Technological Society, said:

> Historians of the future may well look back upon the thirty-five years interval between 1973 and 2008 as the most critical of human history. We must face the fact that we are well into a period in which enormous world forces are converging rapidly and threatening to engulf us all. Indeed those forces may well destroy, perhaps forever, our ability to create a world community in which all people have the opportunity of leading free, abundant and even creative lives, divorced from the traditional scourges of hunger, deprivation, and war.

Why did Brown select 1973 as the starting point? "Because that marks the year when a major critical resource, crude oil, was first used . . . as a major weapon of war." The convergent forces he spoke of are rapid growth of population and affluence; increasing demands for food, energy, and raw materials; decreasing quality of the resource base; changing environment, including climate; rapid technological change; the growing gap between rich and poor nations; the increasing danger of nuclear war that hangs over all humanity; the rivalry of the superpowers; and the increasing vulnerability of complex industrial societies to disruption by a combination of internal and external pressures.

Modern technological societies are confronted by this vast array of problems. They are interlocked, one with another, forming a large web. Solving any one problem will not necessarily ease the functioning of the whole--indeed, it can often make things worse. This is true because the modern technological world is incredibly complex, interconnected, and interdependent.

The Leontief input-output model of the national economy of the United States classifies industries into about 400 major types and requires data for each of these industries about how much it shipped (or received) from every other industry. For each of the many regions of the country there is an input-output table. The resulting 400 x 400 table contains 160,000 numbers. Each number in an input-output table expresses a dependency of one industry upon another. Transactions between regions and industries represent further dependencies; there are many cross combinations. Countries are dependent upon each other in the same way.

Time dependencies also exist: facilities are built and maintained for future use; material is stockpiled for future use; and people are trained for future jobs. There are locational dependencies as well: men, material, and facilities are moved to new locations, not only on the surface of the earth, but below and above.

While we may easily understand the ins and outs of each small part of this vast network of activities, the problem is how to track all the interactions at once. We know that the powerful forces of population growth, shortages of raw materials, food, and energy, growing affluence, and so on, are rapidly reshaping this complexity. There is a fear, based on reasons I will outline later, that the structure interconnecting these activities may not survive these stresses. We see the possibility for all kinds of system failures if we let the changes go on uncontrolled. Is there some way to control this complexity, to steer and reshape it into a more resilient, less interdependent economic system?

The greatest hope for handling the dynamics of change in our complex technological society in the critical years ahead lies in the use of mathematical models and computers. Models are not the solution, but without models, there may be no way to plan a smooth transition.

Having decided that it is unlikely we can survive without models, let us turn to Moses' worry about whether we can survive with models. Suppose we have a model with hundreds of equations and thousands of variables--how do we know if some gross error has been made; what about the many uncertainties, the use of plausible but unverified behavioristic assumptions, and so on.

Large models like the Department of Energy's PIES models can suffer from system antics the same as any other complex systems do. All the cliches apply: "Things aren't working very well (with the model); if anything can go wrong it will." Lincoln made a number of proposals for increasing the reliability of model output which I will review.

The first and foremost is to have a variety of models constructed and run by independent groups; this is the idea behind the Energy Modeling Forum. If I were the Department of Energy, I would support outside groups to run models that compete with internal government models in order to guard against gross errors and loss of credibility.

A second idea is to perform model assessment--careful documentation and verification of data sources and reasonableness of results. Within limits this is useful; I believe a better approach for detecting error is the constant cross checking of results from different models. This could, in addition, increase the credibility of DOE models that represent the official position of the government.

A third idea is to get opinions from industrial experts on whether the results look reasonable from each industry's point of view.

A fourth idea is to have simple as well as complex models. For example, one could have a national energy/economic dynamic model that does not have regional detail. Information from this model could be used to provide national totals to a model like PIES that has regional detail but does not have time dependencies. A national model like PILOT (developed by my group at Stanford) has a very simple set of import-export equations relating the U.S. to the rest of the world. It would be better if there were a hierarchy of models, starting with a grossly aggregated time-dependent international model of the major regions of the world. These could be used to adjust the import-export relations of the U.S. model and, conversely, the national model could adjust the general production functions of the international model. Moving down the hierarchy, next would be a national model with aggregated industries. Next are industrial models more detailed but dynamic. Finally are models studying a particular year, such as 1985 or 2000. These would be non-dynamic national models with regional detail by industry. At the lowest level would be detailed input-output type models for each region.

The problem of the growing disparity between rich and poor nations, the problem of changing our industries so that they will be less vulnerable, the problem of adopting new technologies as traditional sources of resource supply disappear, and the problem of changing and reworking the design of our cities, can all be addressed using models since models can result in balanced and consistent plans of action. We live in rapidly changing times, and we must demonstrate to policymakers and politicians that their respective countries--and the world--can no longer allow them to make policy about complex systems (such as the economy or the environment) without the use of such models. The Energy Modeling Forum and the International Institute for Applied Systems Analysis are part of this educational process.

Much work needs to be done, however, to educate policymakers. That this is so can be seen from a recent survey of key personnel in Washington, D.C., involved in policy making. The following is a paraphrase of some opinions (the first from a "good" Assistant Secretary, the others from Congressional Staff) about modeling and analysis:

> "A good Assistant Secretary knows when to use a number. He knows when a number is good and knows where and to whom to go to get a good number."

> "PIES (the main planning model of the Department of Energy) is irrelevant to anything we do around here. No one, and I mean no one, takes it or any other model seriously."
>
> "The only influence or impact that modeling has is on the incomes of the modelers."

The situation with regard to use of models for policy analysis, however, is not as gloomy as these answers seem to indicate.

Another survey, this time of federally funded modeling projects, concluded that no more than one-third of the models developed achieved their avowed purpose of direct application to policy problems. I regard a score of one-third as very good indeed.

In 1978 the State of Texas acquired from the U.S. Department of Energy the PIES model in order to make its own analysis of what U.S. energy policy should be. This is the first instance I know of in which two rival political groups used models to advance their own positions. In my opinion, this may be the most important single event in the effort to bring about the acceptance of models as the principal tool for analyzing complex issues.

The methodology for using models and computers for formulation of policy exists and has been tested. Model use by policymakers so far has been very limited (at least in the United States). If we accept the thesis that our society is in danger because our technology is designed too tightly, then it is important that models assume a key role in developing plans for moving smoothly to a more resilient technological society. Time may be running short.

Bart Holaday

3. Conditions for Effective Model Utilization in Policy Decision-Making

Introduction

My subject is the impact of models and quantitative analysis on the policy decision-making process, and, more specifically, the conditions that existed when such models and analyses directly affected policy decisions. I will base my comments primarily on my experience in Washington during the period 1968 through 1976. I will also comment briefly on the use of oil and gas supply models in the petroleum industry.

I must make one important disclaimer before discussing conditions that existed when models and analysis affected policy decisions. The nature of the topic is subjective; therefore, everything I say is highly subjective. In any policy decision there are many inputs and many decision-makers. Matters such as energy policy involve numerous House and Senate Committees and nearly every Executive Department. In the Washington policy arena, who ultimately makes the decision, let alone how to weigh objectively all the different input is, in most cases, impossible to know. Any estimate of the precise impact of a specific quantitative model must therefore be highly subjective. Certainly, any policy decision-maker or policy decision group receives input from a wide variety of sources. Often he, she, or they receive analysis and evaluations of alternatives from several different modeling efforts. Even when we could identify a single decision-maker, he or she might not be able to tell us in a specific instance how important model inputs were in the decision process.

With the subjective nature of my comments understood, let me state an even more subjective conclusion. This conclusion is that models and quantitative analysis have seldom had much impact on policy decisions in Washington. For the most part, Washington policy decisions, both in Congress and in the Executive Department, are made in the context of competing and conflicting political interests. I certainly would not argue strongly that it should be otherwise. I also reach a similar conclusion with regard to the use of oil and gas supply models in investment decision-making in the oil industry.

There have been, however, a few times when, in my judgment, the context for the decision or the decision itself has been shaped by modeling efforts or quantitative analysis. At least five important examples arose during my time in Washington. Conditions for effective model use existed in the following five examples:

(1) The Defense Department under McNamara and Clifford--Systems Analysis set the framework for the Defense decision-making process through the introduction of the planning, programming, and budgeting system.

(2) The National Security Council and its staff operations under Kissinger.

(3) The Environmental Protection Agency under Ruckelshaus.

(4) Elliot Richardson's tenure as Secretary of Health, Education, and Welfare.

(5) The period 1974 through 1976, when the Federal Energy Administration and Congress debated various energy legislation and both used the PIES model to evaluate alternative policies.

Conditions

During these five times, similar conditions existed that made it possible for the models and their output to be an effective and instrumental part of the decision-making process. I should state, however, that none of these conditions is necessary nor are all of them sufficient. I am also certain that exceptions can be given to each of the conditions that I will suggest. With that in mind, let me specify the conditions common when models were effective in decision-making processes.

(1) Technical Quality

The models and analysis were technically sound. In most cases, the analytical technique used was close to the forefront of academic thinking at the time it was utilized in Washington. There were also close links between the academic community and the modelers and analysts in the Federal Government. These opportunities usually attracted a high-quality, well-educated, and technically competent core of people who advanced the analytical work.

(2) Focus

The models and the analysis quantified answers to specific policy questions, and they estimated the impact of alternative ways of answering those questions. I have seen, as others have, numerous models that are technically sound and intellectually interesting but that do not analyze any questions of policy interest. For models to have an impact on policy, they must aim at quantifying answers for the questions that policymakers are facing. There is little use in Washington for a model in search of a problem. Although various government agencies have at times funded such efforts, their impact on any policy issues of the time is nil.

(3) Timing

The timing of the analytical or modeling effort was right. By this, I mean simply that the issues required immediate decisions for which the models provided useful input. Moreover, the timing was right in the sense that the models were responsive and provided a quick turnaround time for alternatives being proposed and evaluated quantitatively.

(4) Credibility of Output

The credibility of the modeler and the modeling group with the decision-maker was high. This credibility is based primarily on a proven track record of having provided relevant and correct answers over some reasonable period of time. In addition, credibility is based on the model reflecting realistic assumptions about the real world in which the policymaker is attempting to make decisions.

(5) Communication of Results

Either the decision-maker is trained in the skills of quantitative analysis and modeling or, more commonly, someone I call a "translator" is involved. By translator, I mean someone who understands sophisticated modeling efforts and can translate the results of the models to decision-makers in terms they can understand and believe. Overly simplifying this situation, the translator can take the language of economics and translate it into some sort of "legalese" (lawyer's language).

These are the five conditions that seem relevant and common to the situations mentioned earlier. I am sure there are other characteristics that other observers of the same events in Washington would articulate, but those five seem to me to be the most important and representative of situations when models and quantitative analysis have affected the policy decision-making process. I am sure also that other examples occurred in Washington during the same period. The cases cited, however, were those in which models and their results were effective in providing a basis for reasoned discussion about policy issues and, by virtue of this, have had a substantial impact on the outcome of the policy discussion.

Impact of Modeling and Analysis on Oil and Gas Exploration Investment Decisions

Modeling and quantitative analysis also have an impact on the investment decision-making process in the U.S. oil and gas exploration business. Here again, I emphasize that my impressions are highly subjective, although John Houghton of MIT and I have been working on a paper for the Energy Modeling Forum that examines this topic in greater detail. My conclusion is that models and their results tend to have little effect on investment decision-making. That conclusion is true with the possible exception of two companies (Shell and Exxon) who appear to be ahead of the rest of the industry in terms of their disaggregated supply modeling. Although all companies perform careful financial analysis of specific exploration prospects based on seismic data and estimates of future oil and gas prices, only Shell and Exxon seem to have the capacity to aggregate prospect- and basin-specific information into a nationwide and worldwide system that can be used to optimize all possible capital expenditures.

Conclusions

During my time in Washington and currently in the oil and gas business, sophisticated, quantitative modeling has had little impact on policy or investment decisions. I do however see a trend toward increased reliance on this type of analysis as a viable and important part of both processes. This trend is based on several facts. First, sophisticated models are gaining increased acceptability among a widening spectrum of people. Data Resources, for example, has done much to increase the acceptability of the econometric forecasts of the national and worldwide economic situation and to make those forecasts an important part of both government and private industry views of the future environment. Second, as more people schooled in these techniques attain positions of responsibility, the demand for quantitative analysis will increase. Finally, as more of the policy decisions involve issues that have so many variables and so many uncertainties that no one person can assess them objectively, there will be greater reliance on sophisticated models to assist in the process.

4. The Case Against Central Planning Research

In this paper I will consider transfer of models in the broader context of the relation between federal policy research and local (state-county-city) policy research. To do this, I will contrast two philosophies, centralism and localism. Finally, I present the case against centralism in the United States.

Centralism is both a policy philosophy and an attitude. The policy philosophy dictates that there is a class of major social problems--energy, nutrition, drugs, inflation, defense--that must be studied at the federal level. Results of these studies are translated into federal prescriptions (regulations and other types of law) which are then imposed on local jurisdictions for implementation. Furthermore, the method of research (such as by use of specific models) may be transferred to the local governments, but at the expense of the local government and with a minimum expenditure of time and dollars by the Federal Government.

The attitude of centralism can be called the "Washington syndrome." When men and women go to Washington to serve their country, they are swept into the syndrome in a remarkably short time. They quickly come to feel that Washington's problems are far more important than local problems because Washington is the cortex of the national brain. Another analogy is that Washington is the master parent with 50 children, none of which deserves parental attention any more than any other. Washington is the consciousness of the nation, and all the locals are "unconsciousness" who occasionally surface to the conscious level when they come to Washington--Dulles, National, and Union Station being the entry points.

Since Washington's problems are clearly the most important of the nation, it is necessary to work long hours on them; thus, the burn-out period may be as short as two years. And when people leave Washington, they go down into the unconscious and, like a past dream, may be very difficult to recall. They may have created a large-scale model, for example, but no one in Washington is likely to ask them how they did it, why, or what its weaknesses are. I once invented a method of testing small arms primers for the U.S. Army Ordnance; no one since has ever asked me any questions about it, although it must appear odd to some inspectors.

Of course, the attitude I've sketched is not universal, nor have I mentioned the plight of the heroes and heroines who serve as local representatives of federal bureaus. My chief interest is in the implicit assumption of centralism, which says that research on overall national policies needs the highest funding, and that funding should diminish as one approaches implementation at the local level ("local" being used here to denote state, county, city, or rural town). In other words, the centralist assumption says that justifying a general prescription, e.g., by regulation, is far more difficult and expensive than the task of enforcing it locally.

Localism, in its strongest form, argues that in many instances it is relatively easy to recognize a general policy, such as, that drivers should not exceed 55 mph, should wear safety belts, should drive in the right lane, and so on. What is far more difficult, says localism, is to determine how these general policies should be implemented at the local level, given different cultures, religious customs, ethnic mix, community life, and so on, in different locales of the country.

As an obvious example, designing implementation of a 55 mph speed limit in rural Maine (where 25 mph is often considered reckless) is a totally different process from designing the implementation of the same speed limit in Florida or California. Almost exactly the same thing happens in education. Implementing universal public school attendance in Iowa may be dramatically different from implementing it in Los Angeles where it may be dangerous for a young child to attend school.(1)

Localism does not assume that local implementation always requires more research than does central policy making; the amount of good research required depends on local conditions, and in many instances is far more complicated than current research methods used to determine central policies. According to localism, it is far better to have the central government act as a facilitator rather than as a regulator. If central government is conducting research on a proposed national regulation, then aggregate cost-benefit for the whole nation is <u>not</u> the appropriate measure of performance, because the aggregate hides how the cost-benefit may be distributed among local districts. We are only gradually learning that inequity may be far more important as an ethical base of policy making than aggregate economic gain.(2)

Hence localism argues that research on national policies should include, and often heavily weight, research into the ethical as well as economic issues that local governments must face if a national regulation is passed. For example, if there is no conceivable way in which drug control can be implemented in the central core of cities or certain towns, then national drug laws need to be modified. Those who believe that strict marijuana regulations should be the national policy, and yet have no knowledge whatsoever of how the regulations can be implemented in Los Angeles and Bolinas, California, must be classified as very inadequate planners. None of this is new, of course. Local officials have been saying much the same thing over and over; the problem is getting Washington to listen.

Finally, I turn to the theme of this symposium, the transfer of a research method--a large-scale integrative model of energy--from the central government to its use by local government. Centralism argues that a large energy model should be used primarily for formulating federal policies with respect to energy production and distribution. Because there are 50 states, and thousands of counties, cities, towns, and rural and wilderness areas, the central government cannot afford to pay much attention to the application of its methods locally. If a local area wants to use a federal model, it must negotiate the transfer at its own risk.

Localism, on the other hand, says that transfer should be one of the top priorities in designing large-scale federal models, or any other research methods developed by central government. Central government should regard transfer as one of its chief responsibilities. Central government should perform quality-control checks, design smooth transfer procedures (including careful documentation), and assist local governments in adapting the research methods to the study of local policy making and implementation.

Localism is based on an attitude as well as a rational argument. The attitude is as old as the Roman Empire, where one can easily imagine how people in the provinces must have reacted to edicts received from Rome. The entire argument is also a part of the ancient debate on centralization-decentralization. It could easily be illustrated by the story of large organizations. I do not claim that localism is "right." I do assert that if its case is ignored in the national debate over policy, then national planning will remain in its sorry state. If I've said anything new at all, it is my emphasis on the human side of the debate: people doing things to people.

Notes

(1) The Attorney General of California recently sued the Los Angeles School Board on this issue.

(2) I have discussed this point at some length in Ch. VI of The Systems Approach and Its Enemies, Basic Books, 1979.

Michael Rusin

5. Problems in Modeling Crude Oil Price Decontrol and Windfall Profits Tax Proposals: A Case Study

Introduction

One particular energy policy issue that has been recently debated in the U.S. Congress is the decontrol of crude oil prices and an associated windfall profits tax. This issue provides an excellent case study of the present and possible future role that large-scale energy and economic models can play in policy debates. I emphasize that the views expressed here are not intended to represent the policy or views of the American Petroleum Institute or any of its members.

The windfall profits tax is clearly an important issue, whether or not modelers make any contribution to the debate. It appears that it will be the largest tax ever imposed on a single industry. In the period from 1980 through 1990 the aggregate value of crude oil production will be approximately $2 trillion, and the additional revenues that could accrue to the Federal Government from decontrol and a windfall profits tax could amount to hundreds of billions of dollars. Windfall profits tax revenues could range, under various tax proposals and other assumptions, from $100 billion to over $300 billion.

In any issue of this size, the consequences, trade-offs, or alternatives to various proposals need to be carefully evaluated. The more precisely the issues can be defined and the consequences evaluated, the higher the level of the debate will be. Good models really can help.

Rather than discuss the merits of the various decontrol and tax proposals, I will focus on three topics: first, what kinds of models could help in policy debates; second, where we are now in our modeling capabilities; and third, what is needed to advance the state of the art to the point where we can competently address the important questions raised.

What are the policy issues? Numerous proposals have been put forth recently about how the U.S. crude oil market should be regulated, de-regulated, or taxed. In general, we would like to be able to say, for any of these proposals, what each of its effects will be, and upon whom. However, this is presently impossible, and we have to narrow our scope somewhat.

I will now give some brief historical background. Crude oil prices have been controlled since 1971. Using existing legislative authority the Carter Administration began gradually decontrolling crude oil prices on June 1, 1979. The decontrol process was scheduled to last through the end of 1981. Along with the expected rise in domestic crude oil prices resulting from decontrol, the Administration proposed a set of taxes on crude oil called the "Crude Oil Windfall Profits Tax of 1979."

In the debates over the various crude oil price decontrol and windfall profits tax proposals, the major focus appears to be on the following hierarchy of general questions:

- What is the size of the pie? In other words, how much revenue is generated by various proposals?

- How is the pie divided? What portion of the revenue goes to Federal, state, and local governments, to oil producers, and to others?

- What happens to U.S. crude oil production under various proposals? Does it increase or decrease, and by how much? How are the changes in crude oil production spread over time? What is the effect of increased U.S. production on U.S. imports, other things being equal?

- How do changes in the crude oil market affect the rest of the U.S. economy? How are employment, income, inflation, and so on, affected? How is the world oil market affected?

- Last is a miscellaneous group of questions that fall under the category of general equilibrium issues. Can the major changes and shifts in the economy be discerned and quantitatively described? For example, who really pays a tax on crude oil? Is the tax shifted, and if so, how, and to whom?

Before discussing how each of these policy issues can or cannot be addressed through the use of models, it is necessary to discuss first a few aspects of the U.S. crude oil market, and then the general nature of windfall profits tax proposals.

Categories of Crude Oil Production

A peculiar feature of the U. S. crude oil market is the large number of distinct categories into which crude oil production is divided. By dint of a series of legislative and regulatory initiatives that began under the Nixon Administration in 1971 when crude oil prices were placed under controls, the crude oil market has been segmented into a dozen or more categories.

It is important for the discussion that follows to note generally what these categories are and how they are defined, because each category can and usually does receive a different treatment under various windfall profits tax proposals. This complicates things considerably. Eleven of the major categories can be defined as follows. These descriptions are approximations and simplifications of the legislative and regulatory wording that defines them. Figure 1 lists these categories and their major features.

Lower tier oil, more or less synonymous with old oil, is generally produced from fields discovered before 1973. Upper tier oil is that produced from fields discovered from 1973 through 1978. Newly discovered oil is that produced from fields discovered from 1979 onward. Lower tier oil is itself divided into five separate categories, each of which will probably be differently taxed.

Under price controls, the selling price of lower tier oil was set at about $6 per barrel, or $5.91 in May 1979. The price of upper tier oil was set at about $13 per barrel, or more precisely $12.91. The price of newly discovered oil has generally not been controlled.

Stripper oil is defined as oil produced from wells that produce very slowly, or at an average rate of ten barrels per day or less. These are generally considered to be high-cost wells, and thus are usually given more favorable pricing and tax treatment.

Marginal oil is like stripper oil in that it comes from wells that produce slowly, but its definition also contains provisions that depend on the depth of the well as well as production rates. Deeper wells are presumed to be more costly.

Tertiary oil is that produced by a number of special or exotic production methods including but not limited to the underground injection of steam, underground combustion using air to heat and then thin the oil, or injection of detergents, polymers, alkali, or carbon dioxide. Presently the volumes of crude oil produced by such methods are small, but the known volumes of oil potentially recoverable by these methods are extremely large.

CATEGORY	DEFINED BY:
Lower Tier*	Discovery Date: before '73
Upper Tier	Discovery Date: '73 - '78
Newly Discovered	Discovery Date: after '78
Stripper →	- production volume per well
Marginal →	- production volume per well; - well depth
Tertiary	- production method
High Water Cut Oil	- water/oil ratio produced
Alaskan North Slope	- geographical location
Naval Petroleum Reserve	- geographical location
Heavy Oil	- specific gravity of oil
Independent	- revenue or income of owner

*Has several sub-categories

Figure 1. Categories of U.S. Crude Oil Production: An Incomplete List.

High water cut oil comes from wells that also produce a great deal of water in addition to oil. At least nine times as much water as oil is generally the cutoff ratio.

Alaskan North Slope oil comes from the Prudhoe Bay field in Alaska, and in particular from the single currently producing formation in that field, the Sadlerochit. Although the wells in this field are highly productive, well costs are high, and transportation costs through the Trans-Alaska Pipeline System are also very high.

The Naval Petroleum Reserve at Elk Hills, California, began full-scale development and production several years ago. Although production volume is significant, government ownership of its production makes its tax status moot.

Heavy oil is just that. It is oil that is dense, with a high specific gravity, originally 16 degrees API or heavier, later raised to 20 degrees API or heavier. Heavy oil is generally more viscous, flows less easily, and is harder to produce or pump.

Independent producers are companies or individuals with smaller production revenues or volumes. A number of proposals, as well as the Senate version of the windfall profits tax bill, grant independents differential and more favorable tax treatment, or even exemption from the tax. The Senate version also prohibited members of the 96th Congress and their families from receiving benefits granted to independents.

As can be seen from these abbreviated definitions, a variety of criteria are used to define crude oil categories. They include:

- date of first production;
- production volume per well;
- geographical location;
- well depth;
- the amount of water that comes out of the well with the oil;
- method of production;
- density of the oil;
- the owner's total revenue, income, or oil production; and
- other criteria.

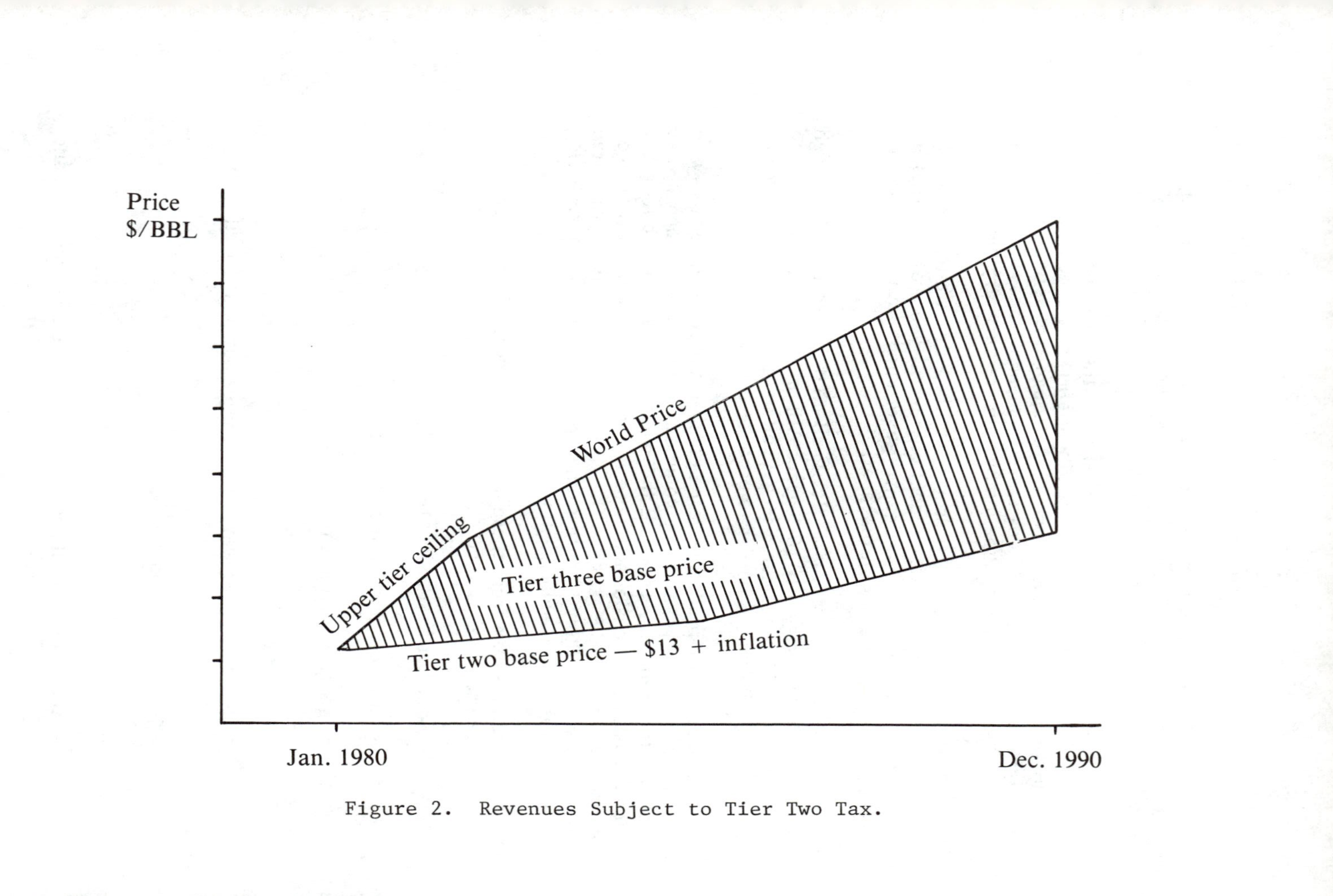

Figure 2. Revenues Subject to Tier Two Tax.

As stated above, because each of these categories can receive distinct price and tax treatment, it must be modeled separately. As we will see below, incomplete data and an incomplete understanding of crude oil production/price relationships render this task rather difficult.

Windfall Profits Taxes

How are windfall profits taxes applied? I will not go into detail, but will give a general idea of how the tax is structured, and show how it is applied to a single category of crude oil.

Each category of oil has associated with it, for tax purposes, a base price. These base prices usually approximate the prices at which that category of oil would be sold if crude oil price controls had been continued. Each category also has a selling price, which may or may not be controlled. The windfall profits tax is applied at a specific, usually flat rate, to the difference between selling and base prices.

As an example, let me indicate how upper tier, or tier two oil, would be taxed. Figure 2 depicts graphically the calculation, over time, of revenue per barrel subject to tax. Upper tier oil, as alluded to above, would include, but not be limited to, production from fields (or properties) that began production in the years 1973 through 1978. First, its selling price would begin at about $13 per barrel in mid-1979 and would be "phased up" to the world or uncontrolled price in late 1981. Second, the base price also starts at about $13 per barrel, but is mandated to rise more slowly than the selling price at the rate of inflation for about six years. After the sixth year it rises somewhat faster.

The difference between the selling and base prices indicated by the shaded area in Figure 2 is likely to grow over time. In the House version of the tax bill this difference is taxed at a 60% rate, and at 75% in the Senate version. With a few exceptions, the tax depends largely upon selling price, base price, and tax rate.

The so-called windfall profits taxes are thus in fact not taxes on profits or windfall profits, but excise taxes, levied directly on crude oil sales, and generally without direct or specific regard for the profitability of oil production.

Why then is the windfall profits tax so finely structured? Why are there over a dozen separate categories of crude oil, each taxed differently? I have no good answers for these questions, except to refer to the historical origin of various categories and their tendency to persist possibly through bureaucratic and legislative inertia. However, many of the crude oil category definitions in fact provide surrogates for cost.

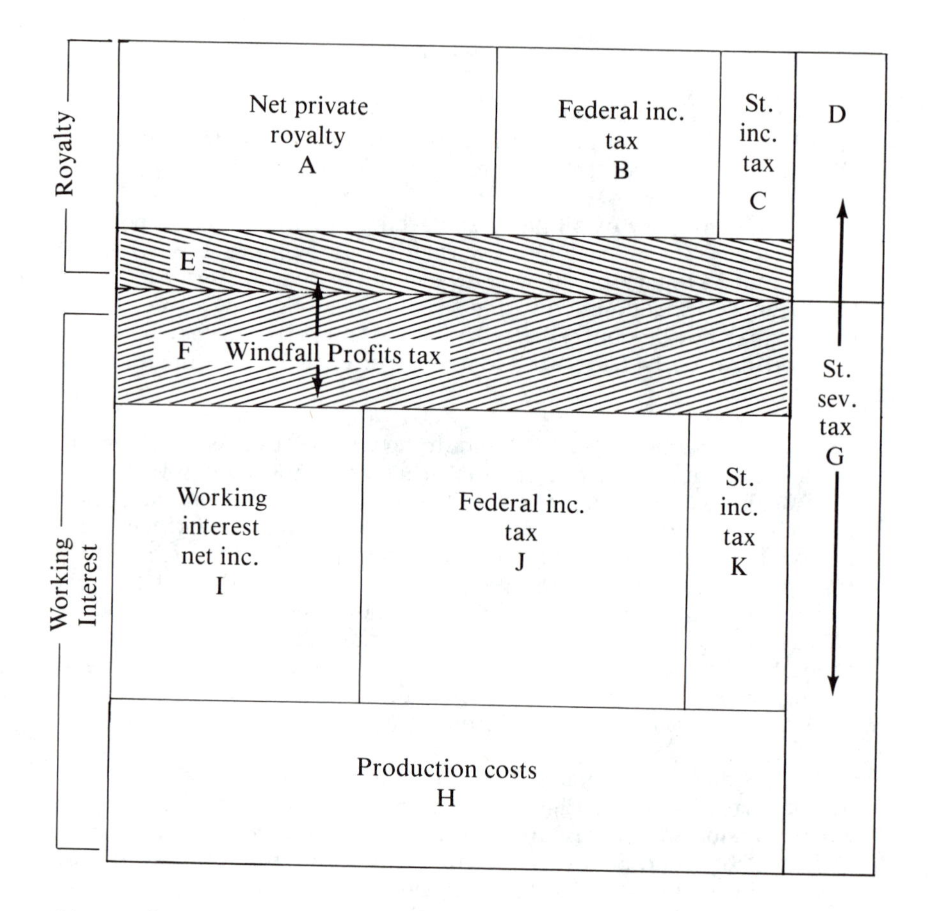

Figure 3. Disposition of Revenues from Sales of Crude Oil Produced on Private Lands.

Some categories such as stripper oil and tertiary oil are generally higher cost. It is difficult to say whether an alternative tax on windfall profits, however they might be defined, would be any simpler or easier to assess or administer.

Time: The time span over which most of the windfall profits tax revenues would be collected is through the end of this decade, or eleven years; thus, our modeling period is most appropriately eleven years. However, the effects of the tax will be felt long after the majority of revenues are collected, and these effects should be considered. In particular, effects on both crude oil production and economic growth are likely to persist beyond the life of the tax. It is hard to assess these effects, but they might be substantial.

Dividing the Pie; Distribution of the Proceeds from the Sale of Crude Oil

When a barrel of crude oil is sold, the proceeds are divided among a number of parties. These include state, federal, and local governments, royalty holders, and the oil producer, or working interest. Royalty holders and producers, in turn, pay state, federal, and possibly local income taxes. In addition, producers distribute a portion of their proceeds to suppliers of inputs to the production process. Precise calculation of windfall profits tax or decontrol revenues that accrue to these various parties depends on accurate specification of how these proceeds are distributed.

We can classify all oil production according to who owns the land from which it is produced: oil produced on private lands, oil produced on State lands, and oil produced on Federal lands. We do so because somewhat different accounting and tax treatment is accorded production revenues from each type of land.

Figure 3 illustrates how the "pie," or proceeds from crude oil sales, is divided for oil produced from private lands. The general scheme is the same for all crude oil categories, but differs in the size of the pie and its pieces. In the figure, pieces of the rectangle are intended to indicate the pieces of the pie, or which parties get which parts of the aggregate proceeds. The size of the block, however, is not intended to correspond to actual allocations.

First, when oil is produced from privately owned lands, the owner of the land usually collects a royalty on each barrel produced. The top portion of the rectangle, including the five blocks labelled A, B, C, D, and E, corresponds to this royalty. The rest accrues, initially at least, to the producer, or working interest, as it is often called.

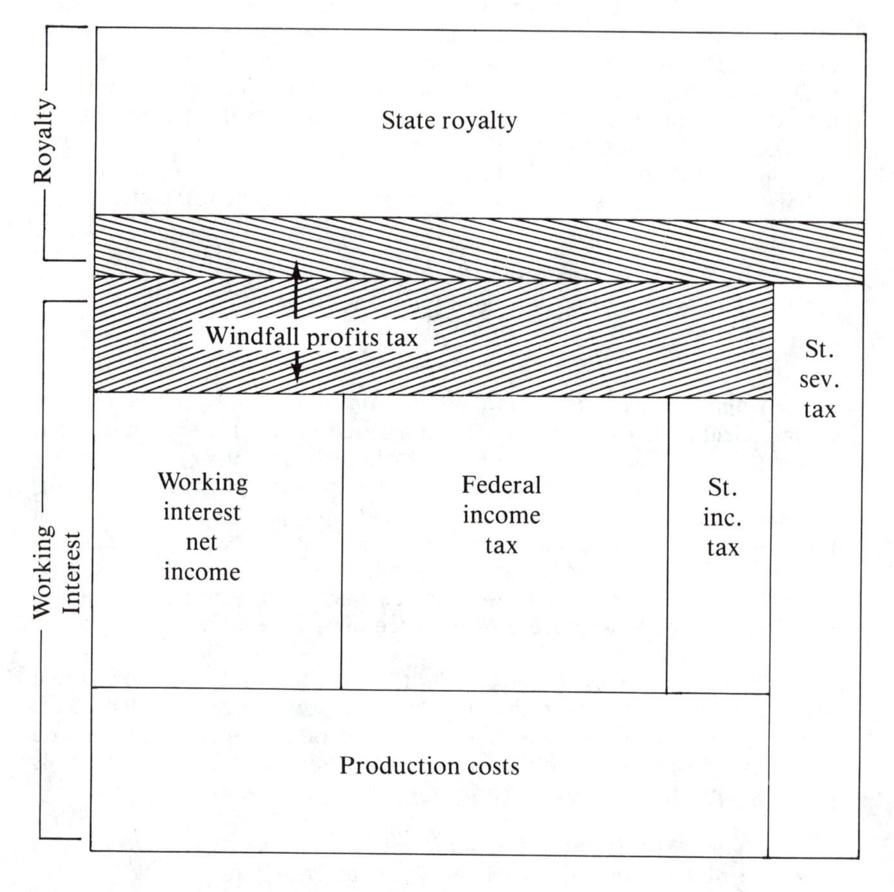

Figure 4. Disposition of Revenues from Crude Oil Produced on State Lands.

Private royalty rates vary widely, but when averaged within crude oil categories that have production from private lands, usually range between six and eleven percent of the value of production.

The state in which the oil is produced usually levies a state severance tax on each barrel of oil produced within its jurisdiction. Severance taxes are applied to both the royalty owner's and to the producer's portions. These two severance tax rates differ but when applied, usually range between two and five percent of the sales price of oil in that category.

Next a windfall profits tax is applied, again to both the royalty holder and the producer. As noted, the amount of tax depends upon sales price, base price, and tax rate. The windfall profits tax on the royalty owner and producer are represented by the shaded areas labelled E and F, respectively.

Royalty owners pay state and federal income taxes on what is left over, as indicated by blocks B and C at the top. What is left, block A, is what they get to keep.

Producers calculate their income tax liability by next deducting the costs of production, indicated by the block at the bottom, labelled H. They then pay federal and state income taxes, blocks J and K, on the remainder, their income before tax, and keep what is left, labelled I, called the working interest net income after all taxes.

State income tax rates are also variable. Average rates among crude oil categories usually range between three and nine percent.

Each of the eleven blocks, or pieces of the pie in the diagram must be separately calculated for each category of crude oil. At best, to be precise, the calculations should be made each month, since some of the legislatively and administratively set prices for certain categories are determined on a month-by-month basis.

The situation for oil produced from state owned or federally owned lands is similar. The major difference is that instead of payments being made to a private royalty owner, royalty payments are made to the State and Federal governments and are taxed in a somewhat different and simpler fashion. Figure 4 illustrates the disposition of revenues from oil produced on State lands.

To calculate the division of proceeds just described, we need a considerable amount of data on prices and tax and royalty rates. Fortunately, although not all the data we would like are available or in the form we prefer, reasonable guesses can be made for the unavailable parts.

Precise determination of tax and royalty rates is important. For example, the aggregate value of domestic oil production during the next decade will be approximately $2 trillion. One percent of this amount is about $20 billion. Thus a "mistake" of one percent in calculating a tax or royalty rate that would be applied to all U.S. production results in calculated misallocation of revenues of about $20 billion, not a trifling sum.

In order to calculate the aggregate dollar volumes, rather than, say, dollars per barrel, we must also know the volumes of crude oil produced, for each category, over time.

Changes in Crude Oil Production

A major portion of the policy debate over price decontrol and windfall profits tax proposals focuses on the trade-offs between on the one hand, the size of the pie and how the pie is divided, and on the other hand, the resulting changes in crude oil production.

As we have seen, before the oil producer sees the net proceeds from crude oil sales, federal royalties, state royalties, private royalties, state severance taxes, windfall profits taxes, and perhaps other taxes as well, are also levied. The cumulative magnitude of these levies can be nearly zero or quite substantial, amounting to well over half the sales price. Taxes and royalties also vary significantly among categories. In other words, tax and pricing proposals affect the proceeds to the oil producer, and are thus bound to affect the volumes of oil produced. Higher taxes can possibly give us lower oil production, and vice versa. The question is how much?

In his remarks to the full Senate at the beginning of floor debate on the Windfall Profits Tax bill, Senator Russell Long of Louisiana, Chairman of the Senate Finance Committee, said:

> Any such windfall profit tax must be a compromise between revenue considerations and the need to provide the proper production incentives . . .
>
> Achieving the proper balance between revenue needs and production incentives requires some difficult decisions. Reasonable people will differ about how best to create such a balance. What the Nation desperately needs, however, is a consensus that will appeal to a broad spectrum of opinion and, thereby, end the divisive argument over oil pricing.(1)

To assess effects on crude oil production, we need a model that relates prices received by producers both present and future, to crude oil production. In fact we need not one model but many. We need a model for each separate category of crude oil, because the multitude of alternative tax proposals affects individual crude oil categories differently.

For example, if it were to be proposed that the windfall profits tax on upper tier oil were to be changed from 60% to 75% we would like to assess the effects on producers of upper tier oil, and their likely production response. This proposal is in fact one specific difference between the House and Senate passed versions of the Windfall Profits Tax bill.

Unfortunately, we are now faced with a state of affairs where the relationships between crude oil prices and production are not generally agreed upon, especially on a category-by-category basis. Models of crude oil production exist, but they are seldom disaggregated to the point where individual categories of oil can be adequately modeled or where there is general consensus on their precision. Newly discovered oil production will change when prices change differently from prices on oil in other categories, such as stripper oil, or oil recovered by tertiary methods, and so on. The need for crude oil production models on a category-by-category basis is a major unmet need, and one which can perhaps be remedied, at least in part, by modelers of the future. What is needed are both better data as well as a better understanding of the production processes and production economics.

We should not lose sight of the fact that oil exploration and production are highly risky and uncertain, and that these risks provide a measure of capriciousness inherent in any model of future crude oil production. Nevertheless, since one of the major trade-offs in the debate concerns the resulting increases or decreases in U.S. crude oil production resulting from alternative proposals, it seems important to address this issue as well as possible. Two additional factors concern the modeling of crude oil production.

First, we need to focus on how money is spent to increase crude oil production. How much will be spent for drilling rigs, production equipment, services, and so on? These sums are apt to be large, and could have substantial effects on the markets for these inputs to the crude oil production process. These effects can be modeled or incorporated directly, or as I will indicate below, accounted for through linkage with a model of the U.S. economy.

My second point concerns the notion that we probably have to calculate two separate sets of numbers: one to account for the effects of the tax laws, and one to account for economic decision making. An analogous situation is one where a corporation issues two sets of financial statements, one to its stockholders to try to

Linkage between crude oil
and macroeconomic models—
a few major variables

From Macro Model to Crude Oil Market Model

Inflation rates/GNP deflator

Real growth rate of imported oil price

Prices of inputs to crude oil discovery
and production

To Macro Model from Crude Oil Market Model

Tax and Royalty revenues

Crude Oil Production/Imports

Demand for Factor Inputs

Corporate Profits and Personal Income

Figure 5.

tell them the state of affairs of the firm, and another set to tell the government, or the Internal Revenue Service, how its taxes have been calculated.

I can also illustrate this point by referring to the distinction between economic profit and accounting profit. Accounting profits can depend upon how a firm finances itself, and generally include normal payments or normal returns to capital. Economic profits do not. Accounting profit determines tax liability, while economic profitability determines business and investment decisions.

As a result of this consideration, proper modeling of the issues raised here probably requires keeping two sets of books--one for determining money and accounting flows, and one for determining economic decision making.

Linkages; Assessing Effects on the Rest of the U.S. Economy and on World Oil Markets

So far we have considered modeling the effects of crude oil price decontrol and windfall profits tax schemes only on the domestic crude oil market. The rest of the U.S. economy and the world have been held constant, so to speak. Clearly the rest of the world is not going to hold still, and parts of it could be quite strongly influenced by changes in the U.S. crude oil market. In particular, it is appropriate to investigate two related and perhaps obvious questions: first, what are the effects on the rest of the U.S. economy, and second, what are the effects on the world oil market, in terms of prices, production, demand, and so on. Other questions could be asked, but these two seem to be most important, and ones where we are now probably in a position to make serious efforts at finding answers.

An obvious way to address these questions is through linkages to a U.S. macroeconomic model and to a model of the world oil market. Alternatively, one large integrated model could be devised, but I suspect that using linked models is cheaper and quicker. Let us first consider linkage to a U.S. macroeconomic model. There are both financial and physical flows, primarily money and crude oil, that must be accounted for. As suggested by Figure 5, changes in price and tax policies within the crude oil sector will produce information on changes in revenues and receipts for a variety of groups.

Changes in crude oil prices, taxes, and royalties are the basic policy variables considered here. They directly affect receipts to federal, state, and local governments. Oil firm incomes are also affected, changing incomes of stockholders and proprietors. Last, increases in resources devoted to oil production will increase

Linkage: Assessing the Effects on the U.S. Economy	
Changes In:	Result in more or less money for:
Crude Oil Taxes and Royalties	Federal government State governments Private Royalty Owners
Oil Company Income	Stockholders Proprietors
Production Costs	Suppliers of equipment and services Employees

Figure 6.

payments to suppliers of equipment and services, to oil firm employees, and so on. Figure 6 lists these important "linkages."

I want to emphasize that the bulk of the changes in financial flows that we are concerned about accrue to or from two categories of economic agents: first, governments at the federal, state, and local levels, and second, those in the oil production sector, including producers, their suppliers, customers, and owners. Under most sets of assumptions, financial flows to and from consumers are relatively unaffected.

We should therefore be careful to verify that the macro model that we use, or link up with, can adequately represent the behavior of these two sectors. Likewise, rates of inflation and the price or supply of goods and services that provide inputs to the crude oil production process are crucial variables that provide linkage in the other direction, that is, from macro model to crude oil model.

Effects on world oil prices, another major linkage variable, can probably only be assessed through use of a model of the world oil market. Let us next consider how this might be accomplished. In the diagram in Figure 7, we start with our major policy variables, crude oil pricing and taxes, represented by the box at the upper left. This set of exogenous variables, along with world oil prices, is then handed over to the model of U.S. crude oil production, represented by the box at top center. World oil prices are also assumed to affect U.S. demand for oil, so that linkage with or incorporation of this effect must also be accomplished. The difference between U.S. demand for and domestic supply of crude oil then, by difference, determines U.S. oil imports, the box at lower right. Along with the estimated effects of world oil prices or world oil supply, the box at upper right, and possibly some inputs from a world oil demand model, not shown, world oil prices are then determined. Of course these effects can be determined either iteratively or simultaneously, depending on the nature of the models used and solution techniques. The scheme used here is only intended as illustrative.

Further Considerations; General Equilibrium; Who Really Pays?

At this point we may still not have enough models or modeling to answer all the important questions. Some of the most difficult questions to answer are those that would be answered by a general equilibrium model, or a model that considers as many shifts, or changes in equilibria as possible.

As an example to illustrate an important effect, consider the question of what happens to, or who really pays, a tax on oil yet to be discovered, specifically, yet to be discovered on federally owned lands. The right to explore for and produce oil on federal lands is

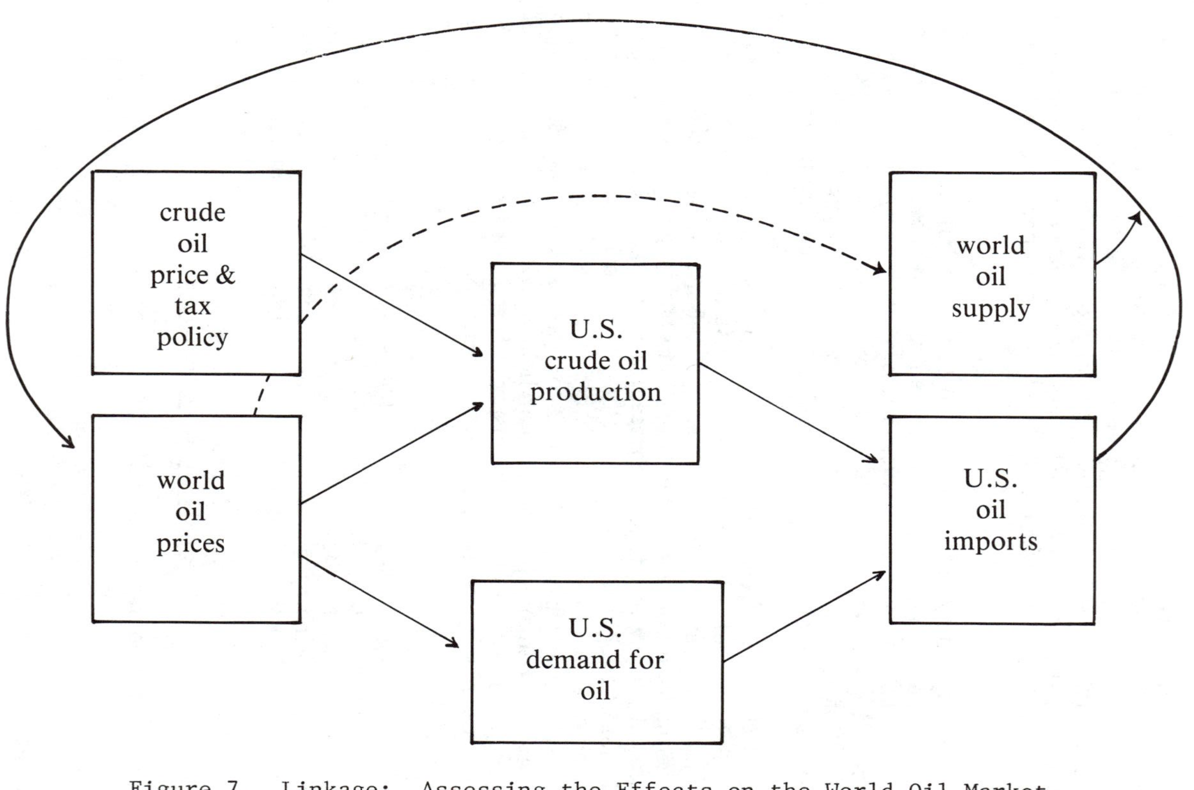

Figure 7. *Linkage*: Assessing the Effects on the World Oil Market.

generally assigned by the Federal Government to the highest bidder in competitive lease sales. On December 18, 1979 the U.S. Bureau of Land Management auctioned off 73 tracts for oil exploration in the Georges Bank area off the North Atlantic Coast. The total of high bids for these tracts was $827 million.

Now consider what would or might happen if the excise tax on yet-to-be discovered oil is increased, by say $1 per barrel. At least two effects take place. First, there will be some oil production prospects which were or appeared to be profitable before the excise tax increase, and are not profitable after the tax is raised. In other words, expected a priori profit was less than $1 per barrel. As a result of the tax increase producers would be likely to ignore these prospects and not produce, or to delay production, possibly hoping for a better tomorrow.

The remainder of the oil production prospects will be somewhat more profitable, that is, worth producing both before and after the excise tax increase of $1 per barrel. What effect will the excise tax increase have? It can be argued that the major effect of the tax increase will be to reduce the amount that prospective producers are willing and able to pay for lease bonuses to the Federal Government.

In other words, had actual or anticipated taxes on oil producers been greater last December, the $827 million in bonuses bid for at the Georges Bank lease sale would have been reduced. The tax in this case takes money out of one of the Federal Government's pockets and puts it in another. In any case, if both the effects mentioned predominate, an excise tax increase could result in a net decrease in future oil production and virtually no net change in total tax revenues.

The task of the modeler is to identify and assess these effects. I might add that a quantitative discussion of the issues just mentioned and others like them has been virtually absent from the policy debate so far.

There are a number of other considerations like these which are important but usually difficult to assess and very often even difficult to identify. I'm not clear how many of these questions can be identified or answered, but it is certainly worth a try.

Conclusions

The proposed windfall profits tax legislation specifies that by January 1983, the President must submit a report to Congress on the effect of decontrol and the windfall profits tax on:

Policy Issue	Required data and Model
size of the pie? dividing up the pie?	distribution of proceeds from crude oil sales; volumes, taxes, royalties
crude oil production? imports?	detailed crude oil production models
the rest of the economy?	world oil market, macroeconomic, model & linkage
who really pays? incidence?	general equilibrium

Figure 8.

(1) domestic oil production;
(2) oil imports;
(3) oil company profits;
(4) inflation;
(5) employment;
(6) economic growth;
(7) Federal revenues; and
(8) national security.

Although the major debate on windfall profits tax proposals may have been completed, with this requirement the period between now and January 1983 still offers a real opportunity to make progress.

Where are we now? Let us look back at our original list of policy issues, restated in Figure 8, and evaluate them in terms of the present state of the art and possible future progress.

First, in determining the size of the pie and how it is sliced under various proposals, we are pretty well off. There may be problems remaining in filling in data cells, but it seems that these can be remedied.

Next, in determining effects on crude oil production, category by category, and the net effect on imports, we are worse off. Future crude oil production is inherently uncertain. Although further progress certainly can be made, we still have a long way to go before achieving detailed crude oil production models of the required precision. We have even further to go to convince policymakers (and others) that these models should be used.

In assessing effects on the U.S. economy, using and linking with macroeconomic models, the present state of affairs is imperfect but workable. Further progress depends upon improved macro models and attention to those sectors where linkage is required. Linkage with world crude oil market models is uncertain. Without progress here in the years to come, we cannot adequately gauge the effects of U.S. energy policies on the rest of the world, and vice versa.

Last, development of usable and generally accepted general equilibrium models seems furthest away. There are not only data problems, but also theoretical and computational problems.

In summary, right now we can do part of the job, but not all of it. Further progress seems likely. Are partial answers or partial solutions still useful? Should modelers be excluded from the policy debates until their works are more acceptable? I think not. Admiral Hyman Rickover, himself a party to numerous nuclear policy debates over the years, once said that "half a brick travels farther than a whole brick." Perhaps half a model can go further, too.

Postscript

On April 2, 1980, the "Crude Oil Windfall Profits Tax Act of 1980," Public Law 96-223, was enacted into law. The law was in most respects a compromise between the versions of the Windfall Profits Tax Act passed by the House and by the Senate. In particular, it contained a provision intended to raise at least $227.3 billion in incremental revenues. To limit revenues to roughly this target amount, the tax would then be phased out, sometime between January 1988 and January 1991, depending upon when and whether the revenue target had been achieved. The final version of the bill (as did the Senate version of the bill) called for a study of the effects of the tax.

On January 28, 1981, President Reagan signed an Executive Order that immediately lifted all price controls on U.S. crude oil, terminating the phased decontrol process already underway.

Notes

(1) Congressional Record, p. S16808, November 15, 1979.

Part 2

Need for Linkage of Micro- and Macroeconomic Models

Russell G. Thompson, A. N. Halter

Introduction: Part 2

Large-scale economic models presently exist for analysis and forecasting of (1) aggregate macroeconomic variables for the economy and (2) detailed microeconomic variables for such sectors as energy, materials, chemicals, and agriculture. Yet no sound means have been developed to link these macro and micro models when, in fact, that linkage should be a prime goal of modelers. Such a gap means that our models are incomplete and less accurate than modern methods should allow. Today, economic relationships are undeniably affected by many outside factors that require detailed micro models for evaluation; hidden costs of regulations (shadow prices) are a case in point. By not incorporating this linkage into our present-day modeling, we are ignoring significant microeconomic consequences on the macroeconomics of the economy, and feedbacks on this aggregate of its component sectors.

The inadequacy of present model linkages is becoming a topic of increasing concern. Documentation of this is found in the Mid-year 1979 Report of the Joint Economic Committee of the U.S. Congress and also in a Texas Energy Advisory Council study of the linkage between the DOE's PIES Model and the DRI macroeconomic model. The Economic Analysis Division of the U.S. Environmental Protection Agency is also in accord with this finding: EPA's macroeconomic evaluations are basically independent of their microeconomic evaluations; and the industrial microeconomic evaluations are done separately, industry by industry. The inadequacy of such a system is clear: it does not consider the economic interactions of producers and consumers in determining prices and their aggregate effects on the economy nor the feedback responses of the economy on its producers and consumers. Without such considerations, no accurate price forecasts can be made and, in turn, no accurate economy-wide impacts can be forecast.

In the current vernacular, "top-down" and "bottom-up" labels are used to describe different modeling approaches. The top-down approach, as used by Klein, begins with a macro-model and adds "satellite" systems to represent the micro side. The bottom-up approach, as illustrated by Heady, begins with a detailed model of an individual sector (here, agriculture) and adds a macroeconomic simulation model. Thompson's integrated industry model, which is also bottom-up, gives the energy/materials sector coefficients of an input-output transactions table as one of its outputs.

These two methods must be integrated into a full feedback model, however, if we are to achieve truly resonant and articulate results. It is not at all clear that economic equilibrium theory is sufficient for establishing existence, stability, and uniqueness under conditions of variable coefficients due to structural changes from external economic forces. A full feedback model must be developed to incorporate economic variations due to government regulations, prohibitions against certain developed technologies, and advances due to new technologies.

This synthesis of microeconomic and macroeconomic models will facilitate many improvements in present-day forecasts. Engineering practice may be imbedded explicitly into the economic concepts to give a future perspective for what is known about technical options and costs.

A well-linked model would have the refinements, capabilities, and advantages of both its components. It would consider the effects of shadow prices on all specific relationships. Further, it would incorporate balanced technical coefficients for economy-wide adjustments in demands, supplies, and prices. Such a model would provide an accurate picture of the national economy and its sectoral components.

Lawrence R. Klein

6. Supply-Side Modeling

Demand Orientation

When an obvious shortfall of aggregate demand in the economy occurred, nation- and world-wide, attention understandably centered on theoretical analysis that was oriented toward explanation of demand--its level, movement, and relationship to potential. Formal demand models paid close attention to the components of demand and their summation, represented by GNP. It is an overstatement to say that the models generated in late 1930 through 1960 looked only at the demand side of the economy. They all had aspects of general equilibrium analysis, much of which is devoted to discussion of the balance between supply and demand in the economy at large.

The demand emphasis comes from the practice of explaining GNP by first explaining its components and then forming, by addition, their total. The components consist of (1) family consumer expenditure; (2) business fixed capital outlay; (3) business inventory investment; (4) public consumption; (5) residential construction; (6) net exports; and (7) government spending on current and capital account.

Technically speaking, total demand is made up of the demand for component parts, but one of these parts represents demand for the purpose of further production. This is definitely true of business demand for fixed capital and partially true of residential construction and public capital formation. The goods demanded will be used in future production and will thus contribute to future supply. In fact, the present discussion of economic policy options focuses to a great extent on private capital formation for modernization and general enhancement of productivity, paving the way for better future supplies. Consumer expenditures on education and training are similarly part of total demand (for services), but also represent investment in human capital. The aim of this kind of outlay is also productivity enhancement.

Contemporary demand, therefore, has elements of supply, but it cannot be said that the typical demand-oriented model is also, automatically, a supply-side model. The demand system elaborates, in great detail, the components of total demand. A demand-side model may go to great lengths to explain many kinds of durable, nondurable, and service expenditures--down to the details of gasoline and oil, furniture, clothing, food, rental housing, entertainment, and such refined categories. Similarly, many kinds of exports, imports, public outlays (federal, state, local), and business capital spending are treated separately. These items should be explained in both real and nominal terms, implying prices from ratios of the latter to the former.

The theoretical explanation of demand, both for components and their total, is mainly academic. The complementary aspect of this analysis is the associated set of policy recommendations, under the heading of demand management. Macro policy formation can steer the economy by operating only at the overall level and not interfering directly with the individual choice mechanism of conventional economic analysis. The control tools are fiscal policy--public spending and taxing; monetary policy; and commercial policy--foreign trade. Public policy makers are viewed as concentrating on these policies at the macro level, "fine-tuning" the economy so that it performs just right--not too vigorously or too weakly.

The successes of these policies in heading the main industrial economies of the world toward full employment for twenty-five years after World War II are remarkable, but perhaps are underappreciated because of relatively poor economic performance during the 1970's. There is much dissatisfaction with current policies of demand management and a search for meaningful supply-side policies. This prompts the present analysis of supply-side model building.

The policy apparatus of demand management is now spoken of, in pejorative terms, as Keynesian economics, and is often dismissed as having been the cause of present inflation with high-level unemployment that is so prevalent now in much of the world. Actually, Keynesian economic policy contributed much to the vigorous post-war expansion of the 50's and 60's, and blocked a return to the depressed state of the world economy of the pre-World War II era, especially in the 1930's. Both deliberate policy choices and the silent implementation of automatic stabilizers were introduced to prevent a return to the depressed conditions of the 1920's and 30's. They function unnoticed whenever recessions or other economic adversity begin to prevail.

A new economic environment, from both the physical and the socio-political sides, and years of nearly full employment, to say nothing of the Vietnam War, led to the breakdown of the era of demand management. These policies cannot be dropped, but they cannot be relied upon to do the enormous job of steering the modern economy through overall policies alone.

In the next two sections, I shall outline the meaning of supply-side components of the total economic model and the associated policy instruments. These are preludes to explicit statement of the new model that tries to accommodate both demand and supply components in one overall model of the system as a whole. It will not be a simple system, as many of the original demand models were. It will first search for what is needed in order to develop adequate analysis of the economy, and then consider whether the system can be simplified. There will be no premise that the final product will be as simple as the familiar multiplier--accelerator model, the IS-LM model, the monetarist model, or any other compact rendition.

The Meaning of the Supply Side

An abstract view of the economy is that it is made up of basic decision-making units--households and firms, not of public sector units, financial sector units, or foreign units.

Households		Firms
demand goods	=	supply goods
supply factors	=	demand factors

Demand models focus attention on the demand for goods by households and the demand for factors by firms. The supply of factors by households and the supply of goods by firms are not treated in adequate detail. It is not that they are completely neglected; it is simply that they are not emphasized or displayed in detail.

Supply of Factors by Households. Labor supply is a principal activity of households and ultimately has an effect on unemployment, which is a key economic variable. Lying behind the relationships of labor supply, we find participation rates for various demographic groups and the whole issue of demographic structure of the household sector.

Participation in the labor force, to be adequately handled, must be separately disaggregated by age, sex, and race. Possibly other groupings are relevant, too, but these are essential. But participation in these groupings must be based on the composition of the groupings. In other words, relationships to explain births, deaths, immigration, fertility, marriage, and health should be part of an adequate supply-side model, unraveling back to demography. Demographic theory and institutions are significant in this respect, but the whole process makes more sense if there is a proper integration with economics of real wages, working hours, working conditions, and labor market conditions.

Labor supply will depend, among other things, on real wages. These will be net of taxes but inclusive of transfers. Tax incentives must be properly included in economy-wide modeling and has been done for some time. Proponents of tax reduction for its own sake argue in the name of increasing incentives, but they are not aware of how much such incentives (or deterrents) are already taken into account for the construction of labor supply relationships. There may be some subjective impulses that are not adequately captured in present relationships, but there are very serious attempts already to reflect this aspect of economic behavior.

The other part of the supply side comes from the production of goods and services by firms. In demand-oriented models, factor demand, in the form of investment demand, is already accounted for. This is not an end in itself because capital goods are demanded in order to produce output on the supply side. Production functions or synthetic cost/supply functions are introduced for this purpose. A proper supply-side model must go beyond the provision of aggregate production functions:

$$Y = F(K, L)$$

where Y = real value added output
K = capital stock
L = labor input

It must go beyond the gross output function:

$$X = G(K, L, E, M)$$

where X = real value of gross output
K = capital stock
L = labor input
E = energy input
M = materials input

Instead of basing the supply-side model on just one or a few strict production functions of the KLEM type, it is better to introduce an entire input-output system to explain the supply side more carefully. The energy and materials (E and M) input values of the production function are taken directly from an input-output system. If they are both measured in constant dollar amounts for (approximate) use in the production function, they make up, together, the column sums of the input-output system.

When modeling the economy as a whole in highly aggregative fashion, we recognize that one sector's (intermediate) input is another sector's output. This is not true of the labor and capital inputs (L and K) as long as the input-output system refers entirely to intermediate flows and excludes capital flows. For the economy as a whole, therefore, it is sensible to use the value-added version of the aggregate production function, but for particular sectors, it is important to use the gross output function because there may not be simple fixed or proportional relations between E and M on the one hand and X on the other.

An input-output system deals with conceptual magnitudes such as X_{ij}, the deliveries of intermediate goods and services from sector i to sector j. The production functions deal with E_j and M_j, the inputs of total energy and materials in sector j, regardless of where they came from.

The production functions are part of the standard macro model of the economy as a whole, but disaggregated by sector. On an economy-wide basis it is well known how to integrate this into a demand-side model (1). On a disaggregated sector basis it is important to combine the Input-Output model with a macro model of final demand, income generation, and market pricing. This system may be called the Keynes-Leontief system, named after the major figures in macro economics and inter-industry economics. In an appendix, the formal structure of such a system is spelled out in detail.

This system was used in the present version of the Wharton Annual Model and for projections, scenario analysis, and economic policy analysis, usually over intermediate (decade) and long-term (several decades) horizons. It is an interactive feedback system. The input-output system of inter-industry flows is driven by the final demand and income generation macro model, but the latter models cannot be solved without knowing the sectoral composition of output and pricing. The two systems must be solved together in joint dependency.

What are the features of this system that will make it important for supply-side modeling? This question will be answered under three headings: (i) capacity constraints and "bottlenecks"; (ii) environment; and (iii) regulation.

(i) Capacity. If there is a shortfall in energy, food, or other resources, economic signals will be transmitted through prices or physical ceilings. The oil embargo of 1973-74 was a striking example of such a shortfall. A combination of the Wharton Annual Model, with the full input-output module, was used together with the quarterly macro model to work out the immediate short run or cyclical consequences (2). At that time, the Wharton Annual Model was ill-equipped to deal with energy detail but has since been disaggregated into more energy sectors--coal, natural gas, oil, types of electric utilities, and types of fuel imports/exports.

Since energy is used in many intermediate processes, the input-output system showed where shortages of other products dependent on petroleum input would develop and where they, in turn, would limit the output of other products. Although energy use was a small fraction of total output (E/X), it had a large impact on overall performance of the economy. This came about through its strategic role and the existence of bottlenecks. It was hard to put these ideas across in 1973-74, but they showed up clearly in input-output analysis.

After the physical limitations of the oil embargo of 1973-74, which caused an actual shortfall of some 2.0 mbd, prices rose, not entirely by market forces, but through exercise of OPEC power. This caused relative prices to change drastically throughout the economy. Prices changed for other products besides fuels, and for a variety of reasons; nevertheless, it was important to have a modeling capability for indicating the effects of these changes on economic performance. This was done by introducing a new feature into Input-Output analysis, namely, to treat the technical coefficients as variables rather than parameters and to explain their shifts over time as functions of relative price changes. The use of the formulas:

$$\frac{X_{ij}}{X_{Kj}} = \frac{\alpha_{ij}}{\alpha_{Kj}} \left(\frac{P_i}{P_K} \right)^{-\sigma_j}$$

explained in the Appendix, show how relative input flows (into the j-th sector) vary with relative price shifts according to the size of the elasticity of substitution, σ_j.

(ii) Environment. Concern with quality of air, water, solid waste, noise, traffic, and other congestion has intensified during recent years. Environmental concerns are usually associated with particular lines of activity such as generation of electricity, production of steel, production of pulp/paper, chemical operations, and many others. Monitoring the level of activity, by sector, in order to be able to estimate particulate waste products or other environmental hazards is a first step. A large multi-sector model like the Wharton Annual Model enables us to generate sectoral output levels for this purpose. It is possible to build a separate model--a satellite model--showing the technical-engineering relationships that affect the environment. Using sectoral activity levels as inputs, the model gives workable methods of automatically linking economic performance with environmental characteristics.

Protecting the environment involves capital and operating costs. These costs go into such devices as stack scrubbers for coal-fired electricity generation, filters for water purification, various disposal devices for waste products, muffling devices for noise, and other capital or operating systems for protecting the environment. These devices should be included in the costs of production of high-quality production, i.e., production that keeps the environment in better condition than would be achieved in their absence. Capital cost variables in the pricing and investment equations of the model should reflect the additional outlays that are needed for protection of the environment. Model scenarios or policy alternatives can be simulated with different degrees and mixtures of environmental measures inserted. They will have to be quantified, and to the extent that they can, the Keynes-Leontief system outlined in this paper is an excellent tool for studying their overall effects, including both direct and indirect influences.

(iii) Regulation. To a certain extent, protection of the environment is achieved through regulation of economic activity, but regulation is a much broader subject that covers many more activities. Regulation of industry may be for occupational health/safety, for controlling rate of return on capital, for meeting efficiency standards, for allocating markets, or for promoting free competition. As in the case of environmental protection, regulation is often extremely worthy, but it is also often costly.

If regulation can be quantified, as in the case of environmental protection, it can be factored into the pricing decision. It may enter as a direct outlay for production or it may affect productivity. For example, imposition of highway speed limits can be translated into an output slowdown for the trucking transportation sector. Accident prevention and fuel efficiency must be taken into account on the plus side, while decreased productivity is allowed for on the other side. As productivity, in the conventional sense, falls, costs and price mark-up will rise. This kind of regulation will manifest itself in higher prices, reduced production, and lower demand. All these direct changes can be traced through the inter-industry sectors of an input-output system to get the full, indirect plus direct, effect.

Increasing attention is being paid to the economic cost of regulation and its contribution to inflation. The kind of model being discussed in this paper seems to be the way of getting at the issues. This is supply-side modeling in every sense of the word.

Supply-Side Policies

Macro models and macro policies of demand management developed, hand-in-glove, together. Accordingly, supply-side models and supply-side policies fit together. Indeed, contemporary economic problems call forth new policies and these new policies need supply-side models in order to interpret their effects.

Supply-side policies are different from macro policies. The latter are not generally directed toward very specific economic activities or sectors of the economy. Supply-side policies, by contrast, are sometimes called structural policies and are aimed at specific issues, specific economic activities, and specific groups.

One of the present ills of stagflation is the persistence of unemployment at levels above what we customarily recognize as full employment. Also we have a maldistribution of unemployment; it is unduly concentrated among some groups in the economy, particularly young people. A structural policy is one that attempts to alleviate youth unemployment. It should make some contribution to overall unemployment through its impact on youth unemployment, but it should be aimed specifically at reducing unemployment among a specific segment of the population. A macro policy would simply try to reduce unemployment or increase employment all around and, in doing so, may generate more inflationary pressure. A youth unemployment program need not be inflationary. A particular structural policy suggested in this respect is one to lower the minimum wage or possibly lower it separately for youths. In a model with well-defined labor supply and demand relationships by demographic groups and minimum wages among the set of explanatory variables, it should be possible to analyze the potential benefits of changed legislation on minimum wages.

Similar structural policies concern changes in unemployment benefits (duration, amount, eligibility criteria) and Social Security contributions. It is conjectured that high unemployment benefits reduce incentives to find work and lower unemployment rates where possible or that high Social Security contribution rates form a cost base on which prices are marked up, thus adding to inflationary pressures, particularly in these times of stagflation.

If the regulatory process is modeled in expanding the supply-side information of a system, then it is possible to examine structural policies of deregulation. It is not a matter of across-the-board deregulation but of selective and gradual deregulation, not only to monitor the inflation content of the regulatory process, but also to study the impact of deregulation on economic performance in the affected branches of the economy.

Overall macro policy on taxation might call for raising or lowering general tax rates on businesses and households in order to achieve GNP or other aggregate targets. A different kind of policy, on the supply side, might aim at adjusting tax rates or other tax parameters so as to stimulate capital formation. The main reason why that is attractive now is that higher levels of capital formation lead to improved productivity and eventually a lessening of inflationary pressure. Thus to get at the fundamentals of inflation from the supply side it is desirable to encourage investment. Tax legislation in the form of investment credits, fast depreciation write-offs, credits for R&D, and favoritism for venture capital are all structural fiscal policies that are expected to work through the supply side. In the first instance, the effects may be most strongly felt on the demand side, but eventually they should show up on the supply side. An appropriate assessment of capital productivity in the KLEM production function will be strategic in judging the effect of such specific tax proposals.

Although these tax policies cited above are intended to work through the supply side, they were often proposed in the context of demand management of the early 1960's. Their supply effects are more important now. But they can be made more specific and more relevant for supply analysis by pin-pointing them even more, towards such strategic sectors as energy investment, agricultural supply, or environmental protection. Differential credits favoring those kinds of capital formation that are most urgently needed are more structural, as policies, and more in the spirit of supply side analysis. Such fiscal changes are not intended to be varied cyclically. Their supply effects are long term and they should be maintained for long periods.

The principal structural policies that have been discussed are domestic, but many of the problems of the present American economy are international in nature--trade/payments deficits with dollar depreciation. Highly structured commercial policies that seek to promote exports would seem to be called for in the present context. Better market research in foreign areas, higher efficiency in export industries, encouragement of potential new export industry lines, and more favorable financing terms for U.S. exporters are all structural policies in the international trade sector that are wanting at the present time.

An Appropriate Role for Demand Models and Policies

The supply-side model does not supplant the demand side; it only supplements it and rounds out the system so that both sides are modeled together. The same is true of economic policy. Macro demand management policies have not been successful in dealing with many contemporary policies, but they did a great amount of good in other respects and continue to do so. They need structural and other supply-side policies in order to meet a wider range of issues.

The Keynes-Leontief model outlined in equation form in the Appendix shows clearly that the total model requires the integration of relationships for aggregate demand and income generation together with supply-side relations. The system is not decomposable and it does not generally admit a solution to either the supply or demand sides separately. Both must be solved together.

In the case of policy formation, there must be full balance in measures to affect the main aggregates through overall fiscal spending, taxation, and monetary control. Policies directed along these lines cannot be expected to bring the economy into balanced equilibrium. There will still be rising prices and high unemployment in present circumstances. There will have to be policies to restrain energy consumption, enhance energy supplies, improve the distribution of unemployment, maintain agricultural output, encourage exports, and raise productivity.

Notes and References

1. L. R. Klein, The Keynesian Revolution (N.Y.: Macmillan, 1947), appendix.

2. L. R. Klein, "Supply Constraints in Demand Oriented Systems: An Interpretation of the Oil Crises," Zeitschrift fur Nationalokonomie, 34(1974), 45-56.

Appendix: The Formal Structure of Supply-Side Models

There are essentially two ways of studying industrial sectors of an economy from an econometric point of view. One approach integrates the detailed industrial composition of an economy directly into a model of the economy as a whole. This is the Walrasian approach. Another methodology is to build a separate (satellite) model of a sector and estimate specific linkage relationships to tie this model to an overall model of the economy. And we shall see that these two approaches can also be combined.

Input-output or interindustry analysis is a standard method of integrating separate industrial output variables in a nationally consistent way. This approach, in its standard form, does not simultaneously determine the macro components of final demand and value added. Much less does it determine interest rates, prices, wage rates, financial flows, inventory investment, and other general economy variables. It usually proceeds with a fixed coefficient matrix and does not relate the changing input-output structure to varying market conditions.

The well known formula of input-output analysis is

$$(I-A)X = F$$

where A = matrix of I-O coefficients

$$a_{ij} = X_{ij}/X_j$$

X = vector of gross outputs

F = vector of final demands

In its simplest form, this system is defined for a fixed A matrix, real (constant-price) X, real F. It is not a closed system in that F must be known in advance, in order to derive a solution for X.

$$X = (I-A)^{-1}F$$

It is the purpose of this presentation to show how a macro model to explain F can be introduced, how market variables can be simultaneously introduced, how the A- matrix can be made variable, and how several variables for each industrial sector can be introduced.

In order to appreciate the modeling problem fully, it is instructive to examine the accounting structure. Consider a current-price diagram as in Figure 1. In the main (nXn) square array, the interindustry deliveries of output from any sector i to any sector j are tabulated. The typical element is X_{ij}; this is the numerator of a conventional I-O coefficient if it is defined in current prices.

The right-hand rectangular group is the matrix of deliveries to final demand. In this case, final demand is split into categories, C (consumption), I (investment), G (government expenditures), E (exports) and M (imports--entered negatively). The column sums of this rectangular matrix give a row of GNP account entries. The row sum is, in fact, GNP.

$$C + I + G + E - M = GNP$$

The disaggregation of F into columns is not unique; it is only illustrative. In detailed model building, it should be much more disaggregated. The Wharton Model has had more than 40 columns for several (model) generations.

The bottom rectangular group is a matrix of values added by sector. Each column of the entire array gives gross output (X), just as each row also gives gross output. While the row sums are broken up into intermediate and final deliveries, the column sums are broken up into intermediate input and value added. The row sums of the bottom matrix provide a column of national income (or national value added) components.

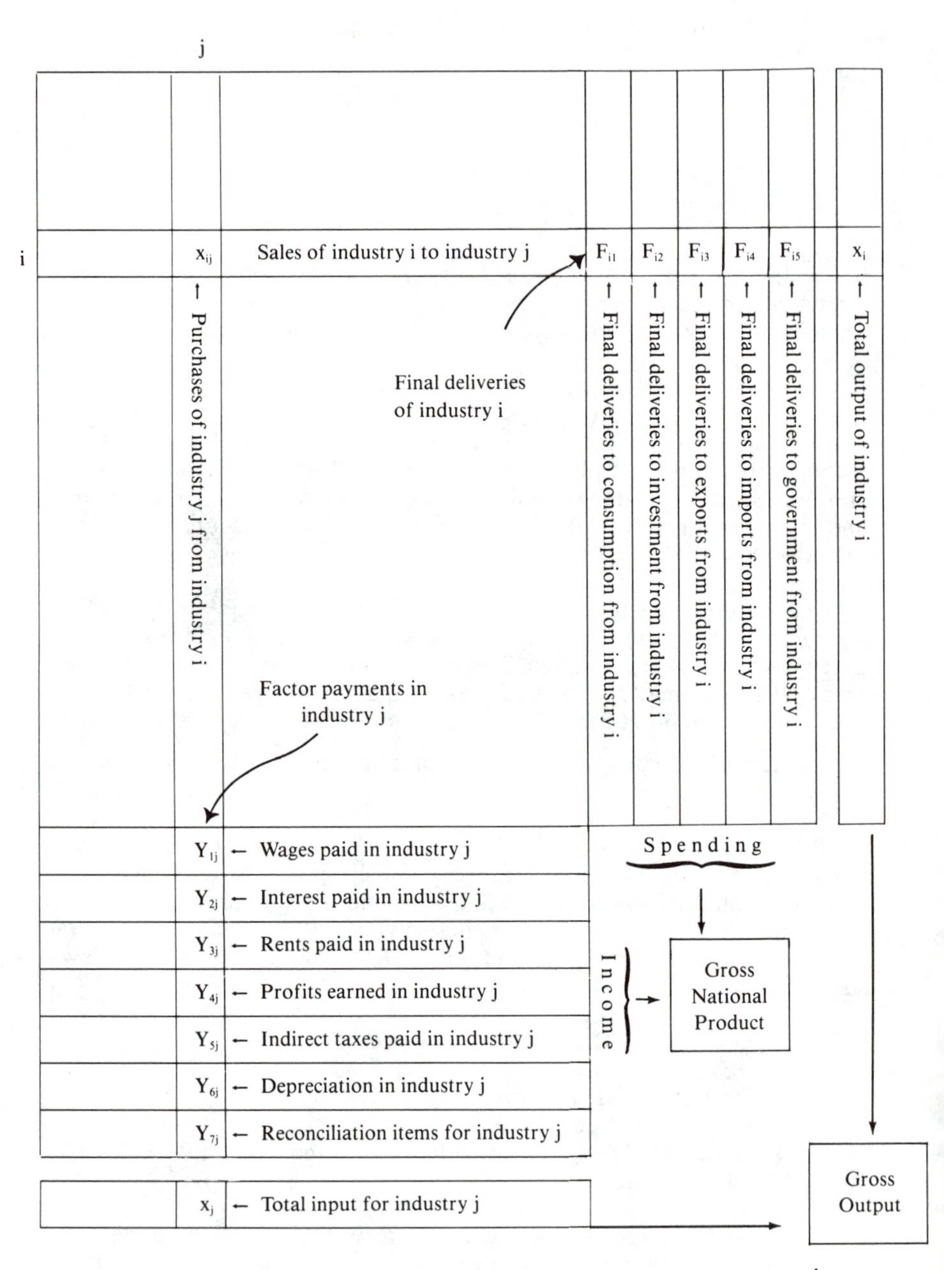

Figure 1. Relationship Between Interindustry Transactions, Final Demand and Factor Payments.

$$
\begin{aligned}
NI &= W + IN + PR \\
W &= \text{wages} \\
IN &= \text{interest} \\
PR &= \text{profits}
\end{aligned}
$$

Value added can be split into more disaggregated components covering royalties, rent, types of wages, etc. If indirect taxes and capital consumption allowance are entered positively, while subsidies are entered negatively, we would have the basic GNP identity.

$$
\begin{aligned}
GNP &= NI + IT + CC - SU \\
IT &= \text{indirect taxes} \\
CC &= \text{capital consumption allowances} \\
SU &= \text{subsidies}
\end{aligned}
$$

Apart from measurement error (statistical discrepancy), these two totals for GNP should be the same, whether measured from F or VA.

Two new matrices will be introduced, one to convert gross output into value added and the other to relate final demand to GNP components. Form the identity

$$VA_j = X_j - \sum_{i=1}^{n} a_{ij} X_j = (1 - \sum_{i=1}^{n} a_{ij}) X_j$$

In matrix terms

$$VA = Y = BX$$

where B is a diagonal matrix with $(1 - \sum_{i=1}^{n} a_{ij})$ in the j-th diagonal location.

Each entry of the F matrix is to be divided by its column total to form a new matrix C. We then have

$$F = C g$$

where g is a vector of GNP components--C, I, G, E, -M in this example.

We can now write

$$(I-A)\, B^{-1} Y = C g$$

$$Y = B\,(I-A)^{-1}\, C g$$

This provides a full linkage between g and Y, between the GNP and the value added, both disaggregated into their own group of sectors.

Macroeconometric models are designed to generate $\mathcal{F}$ (with its components) and Y (with its components). These are national expenditure and income models. These are current-price relationships and, when considered together with physical production relationships, must have properly associated equations for market variables--wage rates, prices, interest rates by sector.

This explains the relationships among elements of $\mathcal{F}$ and Y, with indications about related market variables. It is now in order to deal with variable I-O coefficients a_{ij}. No one argues that the matrix A remains constant through time, but fresh approaches are necessary in order to extend input-output modeling to cover systematically the variation of components of A. This linkage will be developed in relation to the price system; in other words, changes in relative prices will be used as indicators of changes in elements of A.

If the interindustry flows are measured in current prices, we may write them explicitly as

$$\frac{P_i X_{ij}}{P_j X_j} = \alpha_{ij}$$

The <u>conventional</u> input-output measure would be

$$a_{ij} = \frac{X_{ij}}{X_j} = \alpha_{ij} \left(\frac{p_i}{p_j}\right)^{-1.0}$$

If the α_{ij} are stable parameters, then a_{ij} ("real" input-output coefficients) vary as relative prices vary (inversely proportional to (p_i/p_j)). Any two inputs would vary according to

$$\frac{X_{ij}}{X_{kj}} = \frac{\alpha_{ij}}{\alpha_{kj}} \left(\frac{p_i}{p_k}\right)^{-1.0}$$

This result could be derived from an extended Cobb-Douglas production function, where the elasticity of substitution among input pairs is always 1.0, and all intermediate inputs are joint factors in the production function. It is a condition for cost minimization.

A more general formulation is to use the relationship

$$\frac{X_{ij}}{X_{kj}} = \frac{\alpha_{ij}}{\alpha_{kj}} \left(\frac{p_i}{p_k}\right)^{-\sigma_j}$$

For each industry ($j = 1, 2, \ldots, n$) a CES production function instead of a Cobb-Douglas is used to relate output to the joint collection of intermediate inputs. This introduces one more parameter, σ_j, and relaxes the assumption of unitary elasticity of substitution. The α_{ij} are the distributional parameters of the CES specification.

Since σ_j is assumed, in this specification, to be the same between all input pairs into j, we can use cross-industry or sector variation as well as time variation to estimate σ_j.

There will also be labor and capital variation in every sector, so we might write the whole relationship as

Cobb-Douglas $$X_j = A_j \prod_{i=1}^{n} X_{ij}^{\alpha_{ij}} L_j^{\alpha_j} K_j^{\beta_j} e^{\gamma_j t}$$

or

CES $$X_j = A_j \left(\sum_{i=1}^{n} \alpha_{ij} X_{ij}^{-\rho_j} \right)^{\frac{-1}{\rho_j}} L_j^{\alpha_j} K_j^{\beta_j} e^{\gamma_j t}$$

$$\rho_j = \frac{1}{1+\sigma_j}$$

$$\sum_{i=1}^{n} \alpha_{ij} = 1$$

The production function need not separate into intermediate and original factor input functions in a multiplicative specification. Additive specifications are possible as well. The important idea is that a general function is used

$$X_j = F_j (X_{ij} \ldots, X_{nj}, L_j, K_j, t)$$

and the variable I-O coefficients are derived from it using a cost minimization or profit maximization principle. It is this latter idea that leads us to argue that I-O coefficients will vary as functions of relative price variation (1).

This is a static analysis and is most suited for equilibrium models. To the extent that the actual economy, from which data are taken, is out of equilibrium, we need a dynamic adjustment process for short run analysis. This is complicated by the requirement that accounting identities must be satisfied by all data, whether in equilibrium or not.

The treatment of final demand can be consistently implemented within this same framework if complete expenditure systems are used. If consumer demand, for example, can be identified with expenditure categories of the GNP according to sector of origin in the I-O table, we could then use the linear, S-branch, or similar system to estimate all the categories in a way that satisfies budget identities.

Let F_{ic} be the delivery in F from the i-th sector to consumer demand. For each i or subgroup (i, j, k, . . .), we must establish a correspondence with a component of C in the GNP accounts. A linear system would be

$$F_{ic} = a_i p_i + b_i (Y - \sum_{j=1}^{n} a_j p_j)$$

This would enable one to deal with varying ratios F_{ic}/C in the F matrix in the same way that we propose to do this in the A matrix. A similar analysis would have to be extended to other components of final demand, and an extension to dynamic adjustment would also have to be introduced (2).

From an indicative point of view, this is the way that generalized interindustry systems can be modeled on a comprehensive and systematic basis, especially with changing I-O coefficients. Let us now turn to the idea of industrial sector analysis on a satellite model basis.

A model of a sector would be designed to explain such things as production, factor inputs (intermediate and original), factor prices (wage rates, interest rates, material prices, energy prices), output prices, shipments, inventories, profits, and costs. Multiproduct and multifactor processes are possible, as well as stage of process. A typical econometric approach is to write the model as

production function $\quad X_i^S = f_X (K_i, L_i, E_i, M_i)$

factor demand functions

$$K_i = f_K (p_i, r_i, w_i, q_i, g_i, X_i^S)$$

$$L_i = f_L (p_i, r_i, w_i, q_i, g_i, X_i^S)$$

$$E_i = f_E (p_i, r_i, w_i, q_i, g_i, X_i^S)$$

$$M_i = f_M (p_i, r_i, w_i, q_i, g_i, X_i^S)$$

product demand $\quad X_i^D = g (Y, p_i, p_j)$

inventory function	$X_i^S - X_i^D = h(\Delta X_i^D, \Delta p_i, S_i)$
factor price equations	$r_i = k_r(r)$
	$w_i = k_w(w)$
	$q_i = k_q(q)$
	$g_i = k_g(g)$
output price	$\Delta p_i = k_p(X_i^S - X_i^D)$

The production function is a standard relation between output in the i-th sector (X_i) and factor inputs (K_i = capital, L_i = labor, E_i = energy, M_i = materials). The factor demand functions would be specified as a result of an optimization process, ending up as functions of output price (p_i), factor input prices (r_i = capital rental, w_i = wage rate, q_i = energy price, and g_i = materials price). Output level is also a variable in the factor demand function.

The demand function for the i-th sector's product will be a function of price (p_i), national income level (Y), and price of related goods (p_j). Other, more specific, variables may determine demand for the product. In particular, it may not be an issue of general demand in the economy at large, in which case the suitable activity variable would not be aggregate income (Y) but a more specialized variable such as X_j^D or foreign demand, or aggregate production in the economy instead of aggregate income.

Demand and supply need not, and probably will not, balance in every period. Their discrepancy will be the change in inventory stocks. Thus, the equation for (X^S-X^D) is an equation for the flow into or out of stock. Inventory equations are not often very satisfactory, but this one, in the present case, is a very ordinary function, depending on production change (ΔX_i^D), price change (Δp_i), and carrying costs (S = storage plus interest).

The equations for factor prices are made simple functions of corresponding national prices. There may be some specific variable in any or all of these equations, but generally speaking, sector factor prices will follow national factor prices in the corresponding lines of activity.

Finally, output price is assumed to balance supply and demand in this sector; thus p_i should fluctuate in response to its own inventory changes.

Many refinements and variations specific to the sector being studied in satellite mode would, in practice, be introduced, but the main lines of effect would be as indicated. But market variables (product and factor prices) will largely follow national prices. In addition, national demand factors will probably dominate sector i's product demand.

The output of the total national model will be used, after solution, to serve as needed for input into the estimated sector model. It will then be solved for sector variables. These may be used, when appropriate, in the national model. For example, specific intermediate deliveries can be put directly in the I-O system. Other feedback effects can be linked also. The large national model is solved again and input for the satellite model is linked into it. This procedure is then iterated until there are no significant changes on successive rounds. Usually, not many iterations--fewer than six--are needed in order to obtain convergence of both models.

Once the inputs, outputs, prices, and other variables generated by the satellite model are obtained, there are a number of transformations and identities that can be used to obtain measures of such concepts as profits and costs. Costs of intermediate inputs $q_i E_i$ and $g_i M_i$ are aggregative components of the column entries of the input-output table for the i-th sector. By spreading these aggregates throughout the column, it is possible to allow the satellite model to be used to move the A- matrix through time. Of course, types of E and M components can be built into a more detailed satellite model production function and provide more direct information for the input-output matrix. Feedback effects from the satellite can occur in other dimensions, as well. If the demand and output are multiproduct, we can disaggregate into deliveries along the i-th row of the I-O matrix. Satellite models can be used to move rows as well as columns of the matrix. Also, feedback can occur for specific prices, wage rates, final demand, or value added components of the whole large model. If the i-th sector, for example, is the motor vehicle industry, then direct estimates of the automobile component of C (from the GNP accounts) can be obtained from the satellite model for use in the system as a whole.

For interest in any one sector, the procedure of linking a satellite model to a model of the economy, as a whole, is straightforward. If there are several satellite models, they may be interrelated among themselves, and this aspect needs to be built into the specification. Also, care must be taken to ensure consistency with the system as a whole if the collection of satellites makes up a significant and important part of the total economy.

It is probably not feasible to devote detailed attention to all possible satellite models in a system of 50 or more sectors. A preferred approach is to model almost all sectors uniformly through an input-output system. That is the best single approach but any small number of strategic sectors can be accommodated in satellite fashion.

Econometric models are powerful and useful, but they do not provide the only approach to sector modeling nor the best approach for all problems. The use of programming models opens up an avenue for augmenting the strict econometric method and for adding technological information to mainly economic information in a constructive way. A linear program for a sector can be expressed as:

$$p' x = \max$$

subject to $c' x \leq C$

$$TX \leq b$$

where
p is a column vector of output prices
c is a column vector of input prices
X is a column vector of activity levels
T is a rectangular matrix of technological coefficients
b is a column vector of material balances constraints
C is total cost (constraint)

The prices (p, c), optimal (constraint) levels of operation (b), and matrix of coefficients (T) are assumed to be given. The maximization problem can be handled by well-known methods. It is used, however, in a unique way for combination with large-scale models.

From technical information, there are typical programming models for individual industries. Some leading cases are oil refineries, petrochemical plants, steel plants, and electric power stations. For any one of these cases, let us suppose that the details of a linear program are computed for a given price vector.

A pioneer in this development writes:

> This paper explores a new approach to the estimation of a joint production technology. Pseudo data, which are obtained by solving a petrochemical process model for alternative relative prices, are used to estimate a price possibility frontier with three inputs and six outputs.

> Unlike traditional data sources, pseudo data are not constrained by historical price variations, technologies, and environmental controls. As an econometric exercise, the approximation of this process model's detailed piecewise linear production surface by a single equation 'generalized' functional form, the translog, raises a host of interesting empirical and methodological questions (3).

The linear program, as stated, can be solved for given p, c, C, T, b. By varying p and c in a systematic way, we can obtain solutions to the programming problem for each p, c vector. By appropriately varying prices, it is possible to generate and preserve a large body of statistical data on different observations for cost and production, among other variables. The translog specification would make each factor's cost share a linear function (constrained) of logarithms of input prices. By regression analysis on the pseudo data, the equations for cost shares and total cost can be determined. These cost functions may be used for any satellite or directly related cost figures in updated I-O tables. Either the programming problem can be left fully intact for detailed applications, or it can be used to generate pseudo data, to which cost functions will be fitted. These cost functions are then used like any other cost functions that serve for determining intermediate outputs or inputs.

Whether it be through a traditional econometric model or a programming system, the results should be similar and placed in the appropriate position in an I-O table for the express purpose of moving the technical coefficients from one time point to the next.

A priori engineering or operational information is used in constructing the linear program. In this way, new technical processes can be introduced and estimated by econometric cost functions even though observational samples of data are not historically available. That is the power of the method for long-range modeling, as in the case of energy systems.

Sector analysis, whether by I-O methods, ordinary econometric models of satellite systems, or by engineering design, is a promising and growing field of model building activity. All methods together complement one another and no single approach should be relied upon exclusively.

Appendix Notes

1. For specific applications, see M. Saito, "An Interindustry Study of Price Formation," *The Review of Economics and Statistics* 53 (February 1971), 11-25, and R. S. Preston, "The Wharton Long Term Model: Input-Output Within the Context of a

Macro Forecasting Model," International Economic Review 16 (February 1975), 3-19.

2. For total final demand as a single aggregate over expenditure categories, this has been done by Th. Gamaletsos, "Forecasting Sectoral Final Demand by a Dynamic Generalized Linear Expenditure System," Center of Planning and Economic Research, Athens, 1978. He uses a generalization of the linear expenditure system to estimate a consistent set of final demand equations in the context of an input-output system.

3. James M. Griffin, "Joint Production Technology: The Case of Petrochemicals," Econometrica 46 (2, March 1978), 379-396.

Wen-yuan Huang, Earl O. Heady,
Reuben N. Weisz

7. Linkage of a National Recursive Econometric Model and an Interregional Programming Model for Agricultural Policy Analysis

Introduction

Progress in economic modeling and computer capacity provides practical means for analysis of policy alternatives for agriculture and related sectors. Current modeling capability allows construction and use of models that capture the complex relationships of agricultural production activities at the regional level and market activities at the national level. A hybrid model that combines an econometric component with a programming component is an outgrowth of this progress.

This paper briefly reviews the main functions and shortcomings of using an econometric versus a mathematical programming model in policy analysis. Then the paper identifies needs for using linking models and describes a hybrid model recently developed at the Center for Agricultural and Rural Development at Iowa State University. Special emphasis is placed on the linkages between an existing econometric model and an existing programming model.

Traditionally, economists use an econometric model or a mathematic programming model for a positive or normative policy analysis, respectively. An econometric model, which can be as simple as a regression equation or as complex as a system of regression equations, is constructed by specifying the functional relationships of endogenous and exogenous variables according to the economic theory. The functional form is then fitted to historical data through identification, estimation, and diagnostic processes. To estimate possible consequences of a future policy alternative which has previously occurred, the econometric model is the most suitable tool if the economic system remains unchanged. There is an abundance of U.S. agricultural econometric models (2, 3, 4, 5, 7) for answering what was, is, or will be the consequence of any change of the policy.

The model, however, becomes difficult to build and use when historical data are neither available nor long enough. In addition, it is also difficult to build a forecast model which takes into consideration resource, institutional, and other restraints. It is inefficient in analyzing such items as inter- and intraregional competition[1] under a national framework, and interactions between technology levels and resource and institutional restraints. It is impractical, and likely impossible, to develop a regression model that can incorporate spatial problems of the environment and land and water uses.

Although there are many types of mathematical programming models, they possess a common structure, that is an objective function which is to be optimized, and a set of equality/inequality constraints imposed on production activities. A typical programming model is a linear programming model consisting of a linear objective function and a set of linear constraints. The objective function usually is formulated as a linear cost function for minimization problems or as a linear profit function for maximization problems. Technological coefficients used by the production processes and resource supplies available to the production sector are determined through either survey or synthesized data.

The programming model is most useful in estimating the potential regional and/or national production capacity under different levels of production technology, resource availability, and environmental quality consideration. The model frequently is used in policy analysis involving production technologies for which historical observations are not available. It is an effective tool for investigating optimal policy which achieves a defined goal. It also is useful in analyzing spatial problems such as those of soil loss, the environment, water and land allocation, and similar problems. There are also many programming models of U.S. agriculture, illustrated by Heady and Srivastava (8), Meister and Nicol (11), and Dvoskin and Heady (6). Against the stated objective function, these models, in general, can analyze what ought to or would be the consequence of any change of the policy.

The programming model, however, is not designed for structural analysis of past economic systems nor for prediction of future economic systems. Unlike an econometric model, the programming model is not statistically estimated through regression methods using time series data. A trial-and-error method and expert knowledge about the whole production system are required to find a proper programming model to track time series data. Consequently, the model alone is not suitable to describe what has actually happened in the past. The programming model is also difficult to use for prediction purposes. For example, to predict consumers' or producers' responses for a given policy, nonlinear equations frequently have to be used. Existence of nonlinear equations in the model may cause computational problems. Other factors contributing to the programming model being inferior to the econometric model for positive prediction use include the sensitivity of the solution to any change in the coefficient used, and the simplicity of the objective function used in explaining motivation of farmers' or consumers' responses. Greater modeling effort is required to develop a programming model that has the predictive accuracy readily attained with less effort by an econometric model.

Needs for and Cases of Hybrid Models

While there are policy issues that can be best analyzed either through an econometric or a programming model, there are also policy issues that can be best analyzed through a linked or hybrid model that combines an econometric model (or component) with a programming model (or component). Linking the econometric with the programming component allows one component to provide policy variables, information, and analytical structure that could not be specified in the other component. Requirements for a hybrid model can be identified according to intended model uses. The hybrid model (a) can provide detailed and complementary information through the outputs from both econometric and programming models, (b) has a new analytical structure that neither the econometric nor the programming model can provide, and (c) improves the predictive accuracy of an econometric model or transforms a programming model into a predictive model.

The hybrid model belonging to the first group is characterized by linking an econometric (or programming) to a programming (or econometric) model. Output from one model becomes the exogenous values to the other model. This group of models frequently is used for three types of studies.

The first type of study is used to estimate regional production patterns and production related activities (output of a programming model) from a given set of community demands (projected by an econometric model). The econometric component provides information on market activities at the national level while the programming component generates information on production activities and resource use at the regional level. A typical example is used to the NIRAP econometric model (14) with a CARD interregional linear programming model (11). Consumption demands projected by NIRAP are used as constraints in the CARD-LP model to project the competitive equilibrium spatial pattern of agricultural production and resource uses. The purpose of using the model is not to predict a production pattern, but to estimate regional production capability for given national market demands of crop commodities.

A second type of study using output of one model as input into another is done to estimate production potential under different policies or resource capabilities and then to estimate its market impact if the potential were attained. The linear-programming input-output (LP → IO) model used by Sonka and Heady (20) is a typical example. They used the LP model to analyze production impact from policy alternatives. The model's solutions were summarized into ten farm production regions and linked to an I-O analysis to evaluate secondary impacts on rural community income and employment. The model estimates probable market response if the production change is realized.

Finally, a third type of study has used the method to examine the optimal designs or plans in regional production practices to accomplish specific objectives under different national economic situations. For instance, Stockdale (21) used the modified NWA model (11) with the NIRAP econometrical project demands (14) to investigate production activities from several possible alternatives to curb problems related to environmental quality. Numerous potential studies fall in this group. The main purpose of the application is to determine what production ought to occur when under different possible market demands generated from different scenarios for the national economy.

The second emphasis on using the hybrid model is to have a new analytical structure resulting from the combination of an econometric model with a programming model. The model is characterized by the outputs of both the econometric and programming components being determined either simultaneously or recursively. This group of models allows two-way interactions between the econometric and programming components. Particularly, the model allows interface between market activities nationally and production activities regionally. The simultaneous solution model utilizes equations derived from an econometric model as identities (rather than inequality constraints) within the programming model. A typical simultaneous model is the IO-LP model used by Penn et al. (12) for evaluating the impacts of energy shortages on the U.S. economy. Their IO-LP model incorporated input-output data developed by the U.S. Department of Commerce (1974) for 85 sectors into an LP model that contained energy constraint equations. The I-O component in this study provides information of the market impacts due to reduction of energy supplies defined by the LP.

The Quadratic Programming (QP) model with econometrically estimated market behavior restraints also belongs in this latter group. A typical study is the QP model by Meister et al. (10). The model divides the 48 contiguous states into 10 spatially separated consuming regions. Each region is subdivided further into a number of crop-producing regions. The model allows interaction between market mechanisms with regional production activities and resource uses.

Another way to capture two-way interaction between national market activities and regional production is through a recursive modeling structure. At time period t, an econometric (programming) model provides input data to run the econometric (programming) model at the next time period, t+1. The recursive interactive programming models (1, 5, 17, 18) are examples.

The third emphasis of using a hybrid model comes from attempts to improve prediction performance. This model is characterized by including a set of econometrically estimated equations into a programming model to bound regional production responses, or by using an LP model to generate data as input to an econometric model to improve accuracy of prediction.

The recursive programming (RP) model (15) and LP-time series approach (2) are two respective examples. The RP model uses a set of econometrically estimated flexibility restraints to limit the production shift from one time period to another in order to track more closely to observation data. This feature allows the use of a programming model to produce positive estimates. The LP-time series approach uses LP generated output as prior information to an econometric model for estimating crop supply elasticities.

Thus, needs for detailed information, a better analytical structure, or an improvement of prediction accuracy lead to the development of various hybrid models. A hybrid model, according to its emphasis of use, can be identified by one of these three needs. However, these needs are not mutually exclusive; a hybrid model, such as the one we developed, may be built to accomplish all of these needs.

By linking an econometric model with a programming model, analytical capability is extended from a pure normative or positive economic analysis to a wide range of analyses through a combination of unique features available in each of the models. Many positive analyses with prior normative assumptions can be analyzed through the hybrid model. Similarly, normative analysis with prior positive assumptions can be performed.

Several features of the hybrid or linked model have importance for the analysis of agricultural policy, environmental, soil conservation, and land and water allocation. Hybrid models can be used for policy analysis at both regional and national levels: they can estimate regional differential impacts due to a given national policy and estimate impacts on the national level due to a given regional policy. Hybrid models have a dynamic structure, and they can estimate temporal effects of interrelated events. They provide the time frame of various impacts. Their structure relates spatial patterns of supply, resource use, and the technical structure of commodity production to national market structure processes and

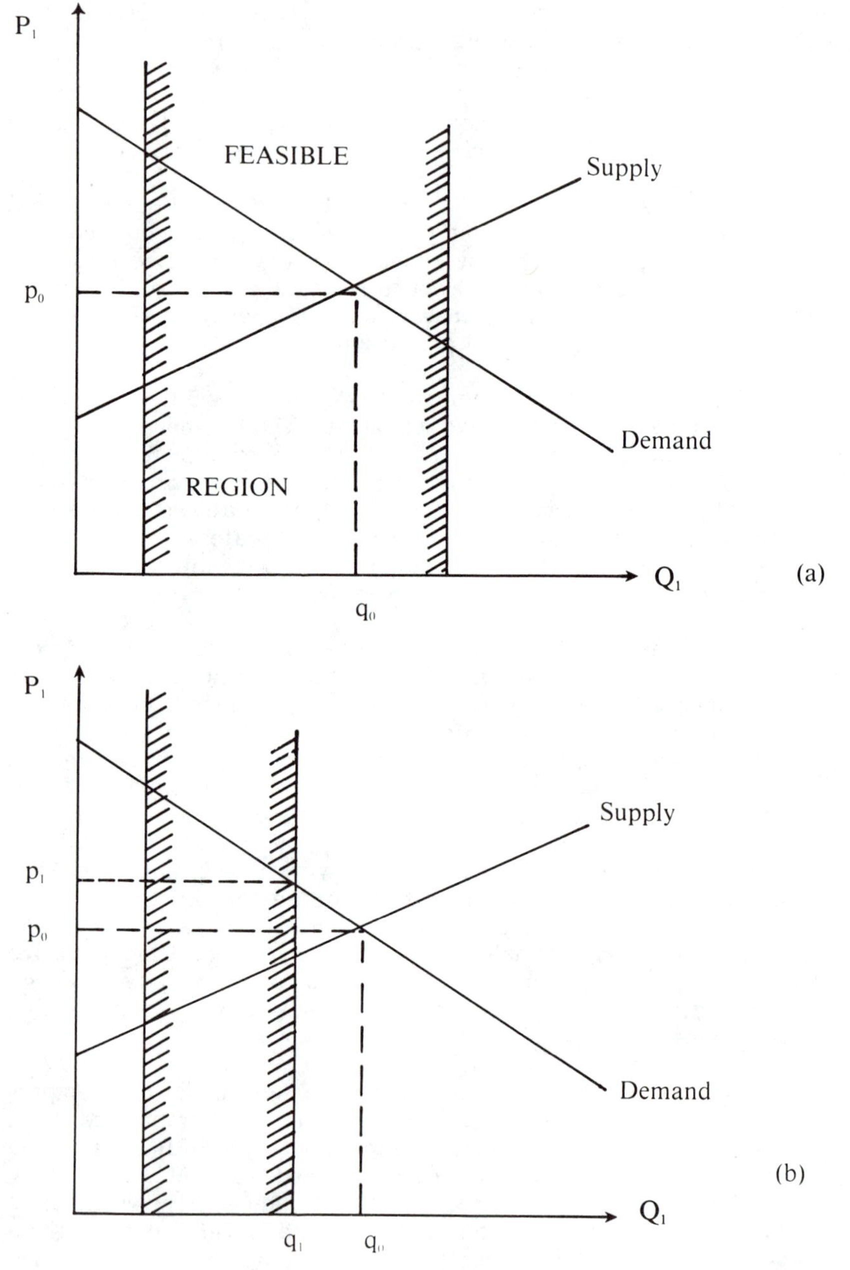

Figure 1. Equilibrium solution (q_0) of econometric component falls in the feasible region (a) or outside of the feasible region (b).

price levels. They can track historical events and give a positive prediction, and can analyze policy issues involving a normative analysis with prior positive assumptions, or positive analysis with prior normative assumptions.

Among possible types of hybrid models, the simultaneous solution model with a time recursive structure appears to be the best approach. It gives solutions that simultaneously satisfy the assumption of both econometric and programming components in each time period. Figure 1 illustrates solutions of the simultaneous model when the equilibrium solution of the econometric component is in or outside the feasible region (dashed lines) defined by the programming component. The model assumes an econometric component consisting of a given set of production restraints and using maximization of consumers' and producers' surplus as its objective function. The model will have the equilibrium solution (q_o) as its solution when the equilibrium solution is inside the feasible region (Figure 1a), and will have a disequilibrium solution as its solution when the equilibrium solution is outside the feasible region (Figure 1b). Since the solutions in both situations will satisfy the assumption of the two components, a consistent result is obtained.

In practice, however, the simultaneous approach is difficult to use. In many cases, both the supply and demand equations may be nonlinear with lagged variables. Under these conditions, computation problems may arise. For this reason, an alternative approach is needed which will give consistent estimates regardless of the location of the equilibrium solution. One of the approaches, the one we use, is a recursive adaptive programming (RAP) model.

Mathematical Formulation of the RAP Hybrid Model

A mathematical summary illustrates the linkage between the econometric and programming components in the RAP hybrid model. Use of the hybrid model as a regional, national, or regional-national model also is illustrated. A simple econometric component and a programming component are used to illustrate the essence of RAP models. Assuming an econometric component consisting of N equations, the hybrid model is expressed as

$$Y_{nt} = \sum_i a_i Y_{it} + \sum_j b_j Y_{jt-1} + \sum_k c_k Z_{kt} + e_{nt} \qquad (1)$$

$$\text{for } n = 1, 2, \ldots, N \quad \text{and} \quad n \neq i$$

where Y_{nt} and Z_{kt} denote endogenous and exogenous variables respectively; a_i, b_j, and c_k are coefficients; and e_{nt} is an error term. Let first I (I < N) endogenous variables be linking variables from the econometric component to a programming component which is expressed as

$$\text{Maximize } [-\alpha_1(\sum_i (V_i^+ + V_i^-)) - \alpha_2(\sum_i\sum_j (W_{ij}^+ + W_{ij}^-))] \quad (2)$$

Subject to:

(1) National production balance restraints

$$\sum_j X_{ijt} + V_i^+ - V_i^- = Y_{it} \quad (3)$$

for i = 1, 2, . . . , I

(2) Regional production response balance restraints

$$X_{ijt} + W_{ij}^+ - W_{ij}^- = \beta_{ijt} X_{ijt-1} \quad (4)$$

for i = 1, 2, . . . , I

j = 1, 2, . . . , J

(3) Production resource restraints

$$\sum_i\sum_j V_{ijl} X_{ijt} \leq R_{lt} \quad (5)$$

for l = 1, 2, . . . , L

where X_{ijt} is the quantity of production of crop i in region j in time period t;

α_1 and α_2 are two arbitrary large constant values satisfying the following conditions: $\alpha_1, \alpha_2 > (P_{ijt} - C_{ijt})$ for all i, j, and t;

where: P_{ijt} and C_{ijt}, respectively, are price and cost of crop i in region j in time period t.

V_i^+, V_i^- represents positive or negative deviations from econometric estimated production of crop i ($V_i^+, V_i^- \geq 0$);

W_{ij}^+, W_{ij}^- indicate positive or negative deviations from econometric estimated projection of crop i in region j ($W_{ij}^+, W_{ij}^- \geq 0$);

V_{ijl} are technological coefficients for using resource l by crop i in region j;

R_{lt} is the maximum amount of resource l available in time t; and

β_{ijt} is the coefficient used to predict the products X_{ijt} from a regression equation which has independent variables such as expected price and other variables.

The objective function in (2) is to minimize the absolute deviation between the programming solution and values of national and regional econometric estimates. Minimizing the absolute deviation is described in Sposito (22). This is the basic structure of the RAP model.

With properly assigned values for V_i^+, V_i^-, W_{ij}^+ and W_{ij}^-, the model can be transformed into a national (Set V_i^+, $V_i^- = 0$), regional, or simultaneous national-regional hybrid model. Adding a cost or profit dimension (Set W_{ij}^+, $W_{ij}^- = 0$) into the objective function, the model can produce additional competitive equilibrium solutions. The CARD RAP model to be described has a net profit function as part of its function.

Figure 2 illustrates the solution $\sum_i X_{ijt}$ and the equilibrium solution in a two-crop model. At each time period, the RAP model will check whether the equilibrium solution Y_{it} is inside the feasible region. If not (A') (either V_i^+ or V_i^- is not equal to zero) the value of $\sum_j X_{ijt}$ (B) replaces the value of Y_{it} and is fed back into the econometric model to adjust the values of all endogenous variables before the RAP model starts the simulation for the next time period (t+1). The adjustment procedure used is defined as the minimum absolute distance approach. It gives the feasible solution which has the least absolute distance between the equilibrium solution A' and the feasible region. Other adjustment procedures can also be used (Huang et al. (9)). When the equilibrium solution is inside the feasible region, no adjustment is required. The value of $\sum_j X_{ijt}$ will be the same as the value of Y_{it}.

CARD RAP Hybrid Model

The CARD RAP hybrid model consists of two components: an econometric component represented by the Center for Agricultural

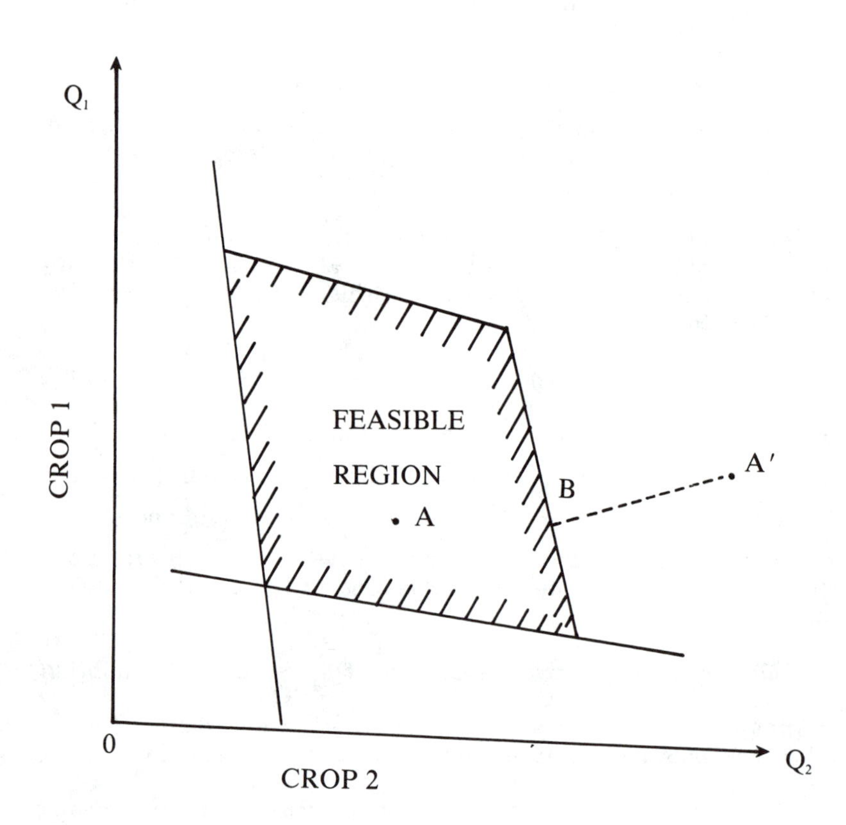

Figure 2. Point B is a feasible solution which has the least absolute adjusted values from infeasible equilibrium solution A' estimated by an econometric component. Point A is the model solution which is the equilibrium solution estimated by the econometric component.

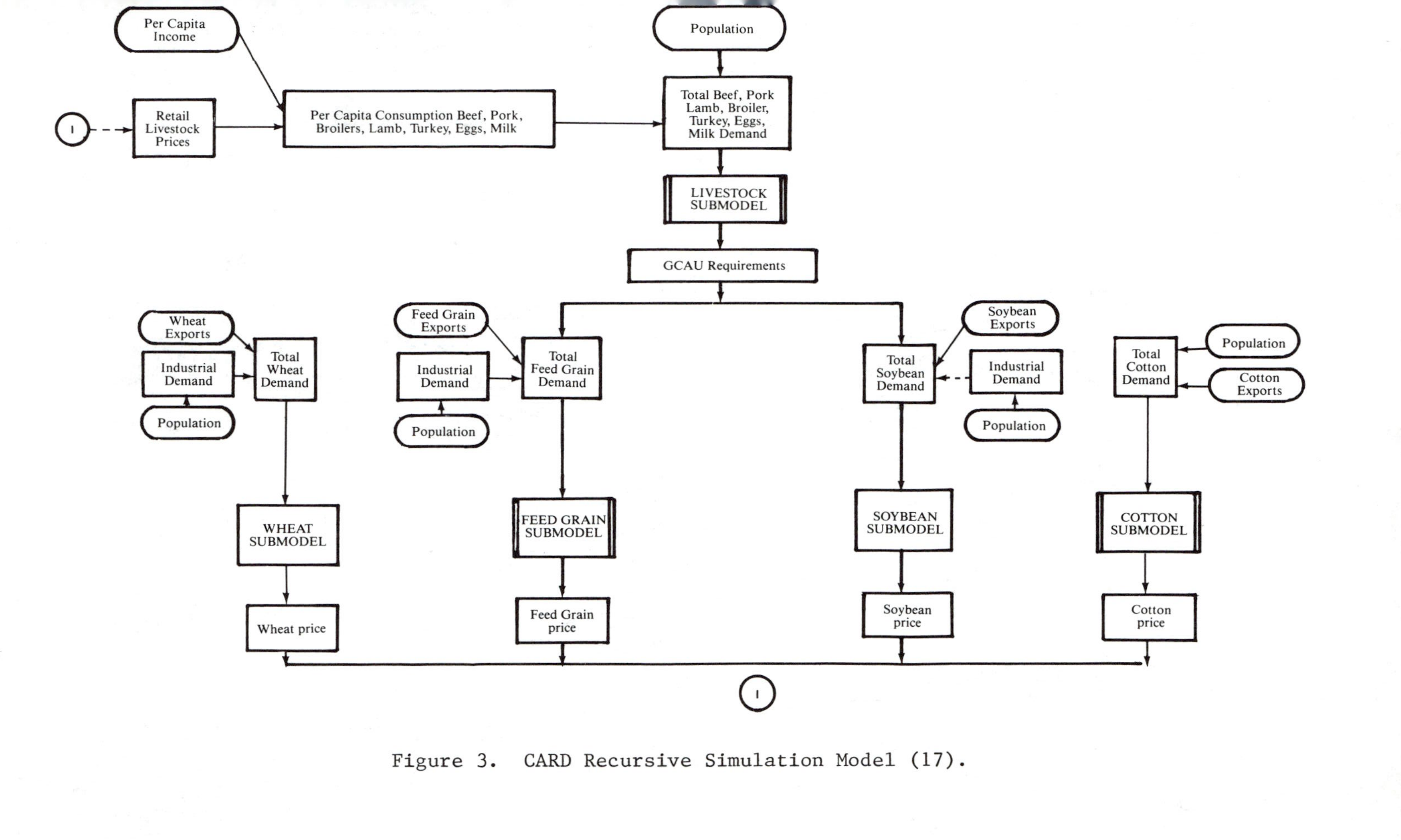

Figure 3. CARD Recursive Simulation Model (17).

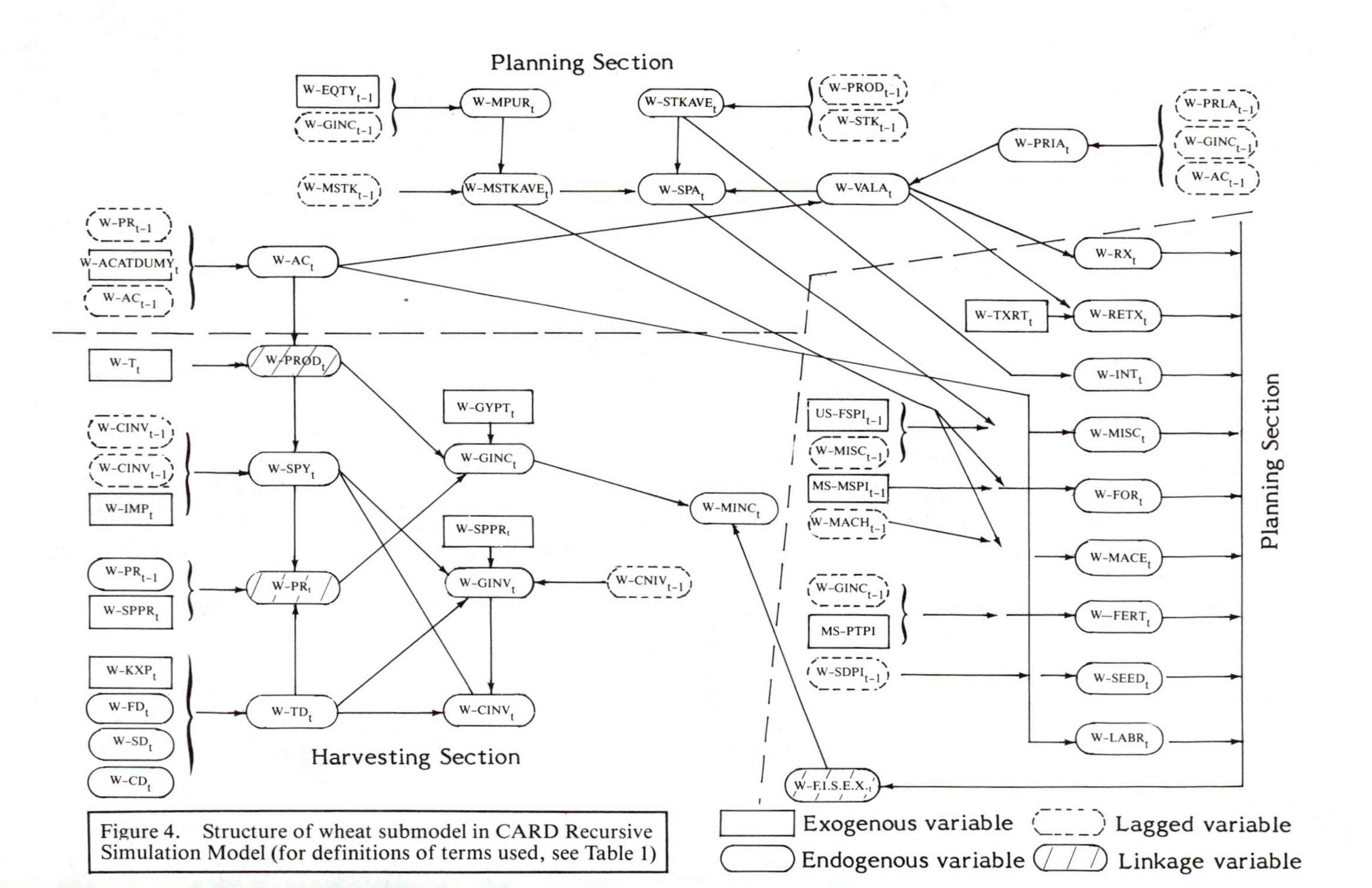

Figure 4. Structure of wheat submodel in CARD Recursive Simulation Model (for definitions of terms used, see Table 1)

Table 1

Definitions of variables used in Figure 4

Planning section		Planting section		Harvest section	
W-EQTY	Equity ratio defined as the value of real estate divided by mortgage debt on that real estate.	W-RE	Real estate expense including interest on land and farm buildings and depreciation repairs and maintenance on farm buildings (million 1947-49 dollars).	W-Y	Crop yield per acre.
W-GINC	Cash receipts and government payments deflated by the implicit GNP deflator (million 1947-49 dollars).	W-TXRT	Tax rate per dollar value of land and buildings.	W-PROD	Crop production (million bushels).
W-MPUR	Machinery purchases (million 1947-49 dollars).	W-RETX	Real estate taxes (million 1947-49 dollars).	W-GINV	Government ending crop year inventory (same unit as production).
W-MSTK	Ending calendar year stock of machinery on farms (million 1947-49 dollars).	W-INT	Interest on farmer-held commodity inventories (million 1947-49 dollars).	W-CINV	Commercial ending crop year inventory (same units as production).
W-MSTKAVE	Average of ending and beginning calendar year machinery on farms (million 1947-49 dollars).	W-FSPI	Index of farm supplies price deflated by GNP deflator (1947-49 = 100).	W-IMP	Crop year imports (same units as production).
W-PROD	Crop production, million bushels.	W-MISC	Miscellaneous expenses including pesticides, small hand tools, binding materials, electricity, telephone, etc. (million 1947-49 dollars).	W-SPY	Beginning crop year supplies defined as the sum of production, carry-in stocks, and imports.
W-STK	Ending calendar year commodity stock on farms (million 1947-49 dollars).	W-FOR	Machinery fuel, oil, and repairs expense (million 1947-49 dollars).	W-PR	Average crop year price received by farmer deflated by the implicit GNP deflator (L, index 1947-49 = 100; dollars per bushel).
W-SPA	Stock of physical assets defined as the sum of STKAVE, MSTKAVE, and VALA (million 1947-49 dollars).	W-MACH	Machinery interest and depreciation (million 1947-49 dollars).	W-SPDR	Average support price levels deflated by the implicit GNP deflator (same unit as price).
W-PRLA	Index of price of land and buildings per acre (index 1947-49 dollars).	W-GINC	Defined in planning section.	W-EXP	Crop year exports (same unit as production).
W-AC	Acreage (million acres).	W-FTPI	Index of fertilizer price deflated by GNP deflator (1947-49 = 100).	W-FD	Crop year demand for wheat as food (million bushels).
W-VALA	Value of farmland and buildings (million 1947-49 dollars).	W-FERT	Fertilizer and lime expense (million 1947-49 dollars).	W-SD	Seed demand (same units as production).
		W-SDPI	Index of seed prices deflated by the implicit GNP deflator (1947-49 = 100).	W-CD	Total domestic crop year demand for all uses, except wheat in which only non-food demand is included (same units as production).
		W-SEED	Purchased plus home-grown seed for individual crops (million 1947-49 dollars).	W-TD	Total demand (same units as production).
		W-LABR	Man-hour requirements (million man-hours).	W-GYPT	Government payments deflated by the implicit GNP deflator (million 1947-49 dollars).
		W-F.I.S.EX.	Production expenses which correspond to the definition used in the Farm Income situation.	W-NINC	Net income deflated by the implicit GNP deflator (million 1947-49 dollars).

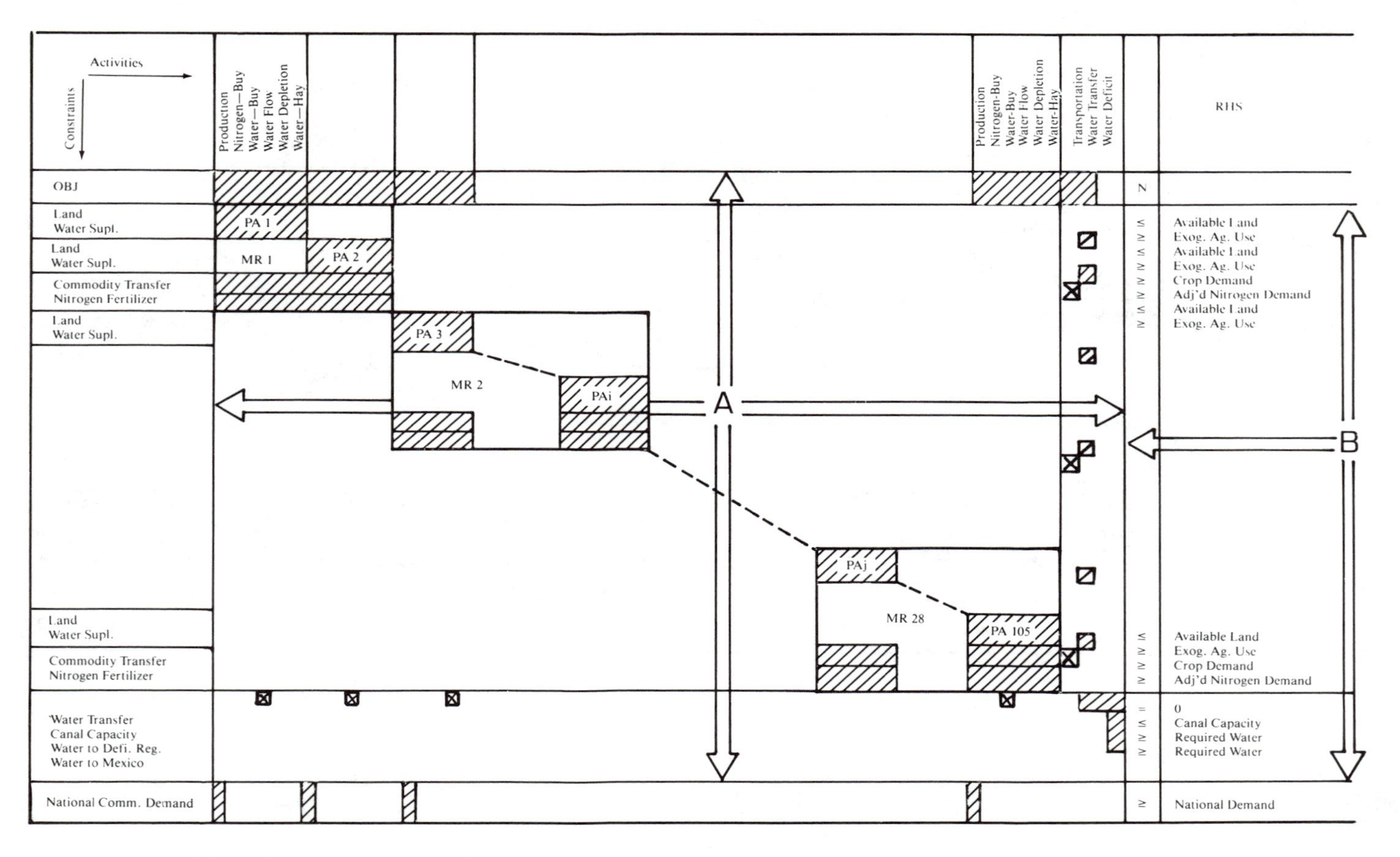

Figure 5. A schematic representation of the structure of the CARD-LP model.

and Rural Development (CARD), Recursive Simulation (RS) model (CARD-RS) and a programming component represented by the CARD Linear Programming (LP) model. These two components, linkage variables between two components, and the feedback processes are described.

Econometric Component: CARD-RS Model

The CARD-RS model, a revised version of the CARD simulation model developed in Ray's thesis (16), is described in Reynolds et al. (17). The RS model is composed of five commodity submodels (Figure 3) representing the major categories of agricultural production. The commodity submodels are used to capture the national production levels of the livestock, feed grains, wheat, soybean, and cotton sectors. Within each commodity submodel, agricultural production is represented by a system of econometric equations. Estimation of these equations was based on yearly aggregate U.S. time series data covering 1930-67.

How the commodity submodel works in the RS model can be best understood by tracing the operation of a typical commodity model such as the wheat submodels. Figure 4 is a schematic illustration of the wheat submodel; exogenous variables are enclosed by rectangular predetermined lagged variables and by broken line ovals, and endogenous variables are in ovals. Definitions of these variables are listed in Table 1.

The commodity submodel is divided into three categories corresponding to the planning, planting, and harvesting decisions incorporated in the production cycle. The planning section determines the levels of such fixed resources as machinery available, new machinery to be purchased, stock of production assets, and the number of acres intended for harvest. Levels of variable inputs such as fertilizer, seed, machinery, and labor requirements are determined in the planting section based on information from the planning section and from previously determined variables. The harvest section provides the production, commodity price, and income estimates resulting from the resource levels committed in the planning and planting sections.

Programming Component: CARD-LP Model

The CARD-LP model (Figure 5), a reduced version of the CARD-NWA model (12), is used as the programming component in the RAP model. It divides the 48 contiguous states into 105 hydrological areas defined during the 1975 National Water Assessment project. Each area is called a producing area (PA). The 105 PA's are aggregated into 28 market regions (MR).

The LP component has six sets of restraints: (a) land re-

Table 2

Potential linkage variables between CARD-RS and CARD-LP components

Output from LP	*As input to RS to determine*
(1)[a] Regional crop production	National production supply and commodity national prices
2. Regional resources uses: crop acreage, water, fertilizer, and other production inputs	Cost of input factors: land rent, irrigation and chemicals
3. Shadow price of each commodity, resource and other institutional regulation	Cost of production, and price of commodity and cost of regulation
Output from RS	*As input to LP to determine*
(1) National and regional commodity prices	Level of resource utilization and regional production response
(2) National and regional production	Values of RHS of production and regional production flexibility coefficients
(3) Cost of production: machinery, labor, chemical and others	Cost coefficient and regional production response
4. Technological coefficients: regional crop yield, water and fertilizer use per acre	Crop yield, water use and fertilizer use
5. Resources supply: land, water and others	Values of RHS of land, water and other rows

[a] Parenthesis indicate linkage variables used in the CARD-RAP model.

straints for dry cropland and irrigated cropland; (b) water supply restraints; (c) commodity transfer rows; (d) nitrogen fertilizer restraints; (e) water transfer and canal capacity rows; and (f) crop production constraints.

Activities in the programming component include: (a) crop production activities, where a crop production activity is defined as a crop rotation (based on the crops corn, corn silage, sorghum, sorghum silage, nonlegume hay, wheat, oats, barley, soybeans, legume hay, cotton and summer fallow) on a specific land class using specific tillage and soil conservation practices; and (b) water use activities of water-buy, water movement by natural flows, and water transfer between supply regions and inter- and intrabasin transfer; other water use activities; commodity transportation activities; and nitrogen-buy activities.

As illustrated in Figure 5, the model consists of 28 market regions, each of which contains one or more producing areas. The constraints in each PA are the land and water supply constraints; activities are production, water-buy, water-flow, water-depletion, and water for hay use (conversion of irrigated hay to nonirrigated hay). The PA's in a market region are related by commodity demand and nitrogen fertilizer constraints and water transfer activities. Interdependent relations between MR's are built through the national demand constraints and commodity transportation activities. Other water constraints including canal capacity, water delivery to deficit areas, and water to Mexico are also shown in Figure 5.

Linkage Variables and Adaptive Processes

The linkage variables are the predetermined variables in one model (component) whose values are determined by the operations of the other model (component). The variables considered for linking the CARD-RS and the CARD-LP models are listed in Table 2. These groups of variables can be used to transfer information from the LP (RS) model to the RS (LP) model. Regional crop and livestock production by commodity can be aggregated as the production supplies as input to the RS model; the regional resource utilization can be used to determine the demand for resources input and their costs. The shadow prices of commodity, resource, and other institutional regulations provide valuable information for determining price of commodity and cost of regulation. The shadow price also gives important information for determining capital investment in the planning section of the RS model. Conversely, the RS model can provide five groups of information to the LP model. The national and regional prices are important factors in determining levels of resource utilization and production response and facilitates formulation of regional production response. Forecasted

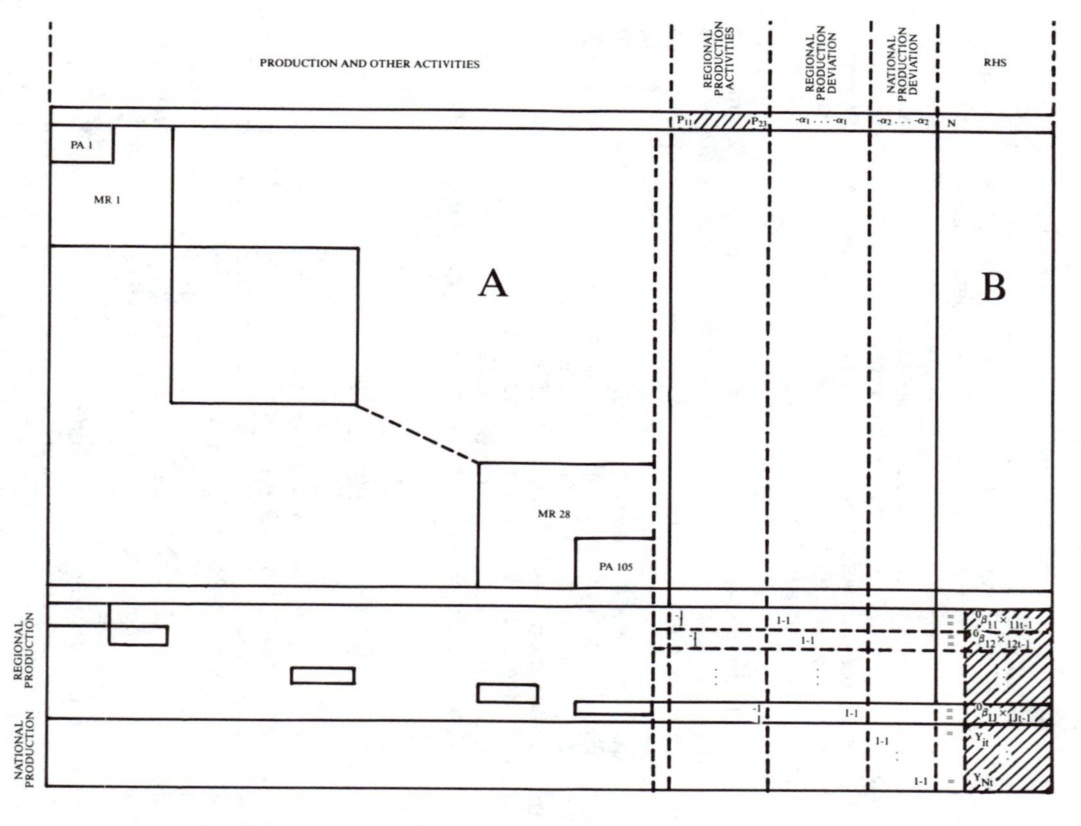

Figure 6. The linear programming component of the RAP hybrid model.

national and regional consumption based on population provides the values for the demand RHS in the LP model. The projected costs of productions provide cost coefficients in the objective function of the LP. Crop yield, water, and fertilizer use can be estimated by the RS and used as technological coefficients in the LP. The supply of resources estimated by the RS is needed for the RHS values of resource restraints in the LP.

Currently, the linkage variables that transfer information from the econometric component to the programming component are national commodity price (PR_t), national production ($PROD_t$), and production expenses (F.I.S. EX_t). The PR_t is used in determining expected regional commodity price (P_{ijt}) which is used in constructing objective function[2] of the programming component, and in determining regional production response coefficients (β_{ijt}).

The $PROD_t$ gives quantity of expected national production (Y_{it}) to be used on the RHS of linkage set restraints. The F.I.S. EX_t provides the basis to calculate the cost coefficients (C_{ijt}) in the objective function. Yield variables have not been used as linkage variables to adjust the yield data because a short-run yield model is yet to be developed.

National and regional crop productions are variables used to transfer information from the programming component to the econometric component for the adjustment mentioned before. Use of resource shadow prices to determine the capital investment appearing in the planning section of the econometric model is yet to be explored.

Figure 6 shows the modified LP component used in an RAP hybrid model. The A and B parts are from the CARD-LP model described in Figure 5, excluding commodity transfer rows. The row in the component consists of the main structure of the CARD-LP model, as indicated by A and B regional and national production rows. In addition to the column of the CARD-LP, the LP column in the figure also includes regional production levels (X_{ijt}), regional production deviation variables (W_{ijt}^+, W_{ijt}^-), and national production deviation variables (V_{it}^+, V_{it}^-). Since the RS model has a feed grain submodel, corn, barley, sorghum, and oats in the LP model are converted into a feed grain row at the national production level. Coefficients of costs, prices, and regional production responses are updated annually by using information generated by the RS model. These coefficients are indicated by the shaded block in the figure.

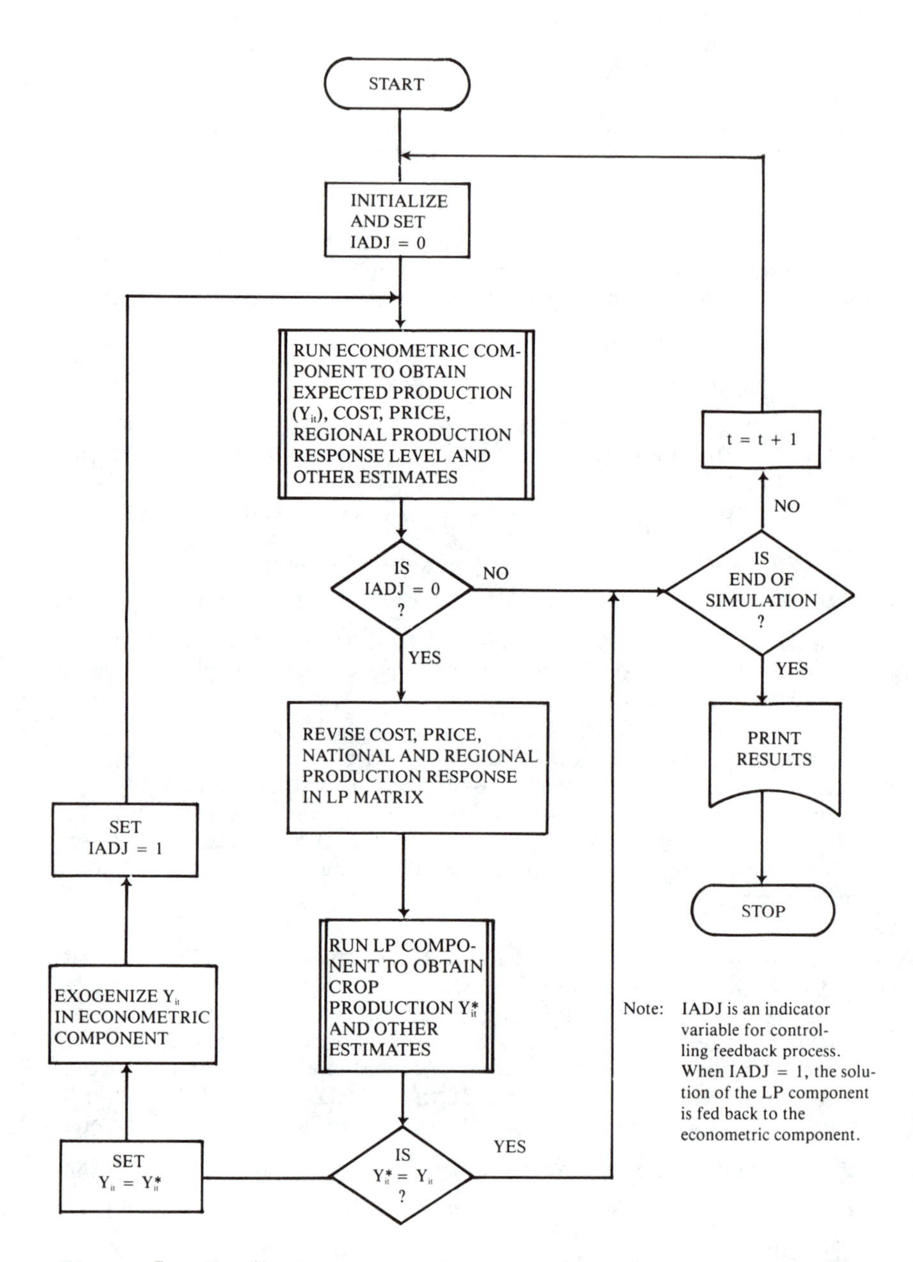

Figure 7. Feedback process in recursive adaptive programming model.

Operation of this hybrid model can be illustrated by the flow-chart in Figure 7. Analysis is initiated by running the econometric component to estimate expected aggregated production (Y_{it}), costs, prices, regional production response range, and other information. These estimates are used to revise the coefficients in the LP matrix. The revised matrix is used for the LP run to obtain the aggregate national production Y_{it}^*. If the values of $Y_{it} = Y_{it}^*$, that is, V_{it}^+, $V_{it}^- = 0$ for all i, the operation moves to the next time period. If $Y_{it} \neq Y_{it}^*$, the hybrid model sets $Y_{it} = Y_{it}^*$ and revises the values of endogenous variables. Part of these revised variables become lagged variables for econometric component to be run in the next time period.

Conclusions and Applications

The CARD RAP hybrid model is developed for analyzing the policy issue which has temporal and spatial impacts. The programming component of the model provides policy variables relating to spatial production technology, resource supply, and production pattern, while the econometric component provides the variables including processing, consumption, market prices, farm income, farm assets, and capital investments. For a given set of agricultural policies, the model can be used to simulate a dynamic sequence of interrelated production and market activities over space and through time with a consistent set of estimates that satisfies assumptions of both components. The model will allow the policy maker to expand its analytical capability from a pure normative or positive economic analysis to a wide range of analyses through the combination of unique features available in each component. Policy issues of positive nature with prior normative assumptions, or normative nature with prior positive assumptions, can be analyzed through the hybrid model.

Currently the model is used to investigate the impact of pesticide regulation on cotton and soybean production. The results illustrate time paths of production, price, farm income, and other impacts on the national level, production shifts between regions, and change of cropping patterns with regions.

Footnotes

1. Lee and Seaver (9) demonstrated that a simultaneous econometrical model can be used to investigate spatial equilibrium of the broiler market. However, considerable computation is required to find the solution. For n regions, $2^{n(n-1)}$ solutions have to be found.

2. A profit function is added to the objective function (2).

References

1. Baum, K.H. A National Recursive Simulation and Linear Programming Model of Some Major Crops in U.S. Agriculture, Ph.D. Dissertation, Iowa State University, Ames, Iowa, 1978.

2. Boutwell, W. et al. "Comprehensive Forecasting and Projection Models in the Economic Research Service," Agricultural Economics Research 28, 41-51 (1976).

3. Chase Econometric Model, CEA80A, Chase Econometric Association, 1977.

4. Chen, D.T. "The Wharton Agricultural Model: Structure, Specification, and Some Simulation Results," American Journal of Agricultural Economics 59, 107-166 (1977).

5. Day, R.H. "Recursive Programming and Supply Prediction," Agricultural Supply Function Estimating Techniques and Interpretation, pp. 108-127. Edited by Earl O. Heady, et al., Iowa State University Press, Ames, Iowa, 1961.

6. Dvoskin, D., E.O. Heady, and B.C. English. Energy Use in U.S. Agriculture: An Evaluation of National and Regional Impacts From Alternative Energy Policy. CARD Report 78, The Center for Agricultural and Rural Development, Iowa State University, March 1978.

7. DRI Agricultural Service. The 1977 Version of the DRI Agricultural Model. Working paper No. 5. Data Resource, Inc. October 1977.

8. Heady, E.O. and U.K. Srivastava. Spatial Sector Programming Models in Agriculture (Iowa State University Press, Ames, Iowa, 1975).

9. Huang, W-Y., R.N. Weisz, and E.O. Heady. Econometric-Programming Hybrid Models for Spatial and Temporal Analysis of Price, Production, and Resource Use: Development of A Recursive Adaptive Programming Hybrid Model. To be published as Technical Report, Center for Agricultural and Rural Development, Iowa State University, Ames, Iowa, 1979.

10. Lee, T.C. and S.K. Seaver. "A Positive Model of Spatial Equilibrium with Special Reference to the Broiler Markets," in Studies of Economic Planning Over Space and Time edited by G. Judge and T. Takayama, 1973.

11. Meister, A.D., C.C. Chen, and E.O. Heady. Quadratic Programming Models Applied to Agricultural Policies (Iowa State University Press, Ames, Iowa, 1978).

12. Meister, A.D. and K.J. Nicol. A Documentation of the National Water Assessment Model of Regional Agricultural Water Use and Environmental Interaction, CARD Report 65, Center for Agricultural and Rural Development, Iowa State University, Ames, Iowa, 1975.

13. Penn, J.B. et al. "Modeling and Simulation of the U.S. Economy with Alternative Energy Availabilities," American Journal of Agricultural Economics 58, 663-671 (1976).

14. Plessner, Y. Quadratic Programming Competitive Equilibrium Models for the U.S. Agricultural Sector, Ph.D. Dissertation, Iowa State University, Ames, Iowa, 1965.

15. Quance, C.L. Agriculture in the Third Century, ESCS, No. 1, May 1976.

16. Ray, D.E. and E.O. Heady, "Government Farm Programs and Commodity Interaction: A Simulation Analysis," American Journnal of Agricultural Economics 140 (1972).

17. Reynolds, T.M., E.O. Heady, and D.O. Mitchell. Alternative Futures for American Agricultural Structure, Policies, Income, Employment and Exports: A Recursive Simulation, CARD Report 56, Iowa State University, June 1975.

18. Schaller, W.N. "A National Model of Agricultural Production Response," Agricultural Economics Research 20, 33-46 (1968).

19. Sharples, J.A. and W.N. Schaller. "Predicting Short-Run Aggregate Adjustment to Policy Alternatives," American Journal of Agricultural Economics 50, 1523-1535 (1968).

20. Shumway, R. and A.A. Chang. "Linear Programming versus Positively Estimated Supply Functions: An Empirical and Methodological Critique," American Journal of Agricultural Economics 59 (1977).

21. Sonka, S.T. and E.O. Heady. Income and Employment Generation in Rural Areas in Relation to Alternative Farm Programs, North Central Regional Center for Rural Development, Iowa State University, December 1973.

22. Sposito, V.A. "Linear and Nonlinear Programming," Iowa State University Press, Ames, Iowa, 1975.

23. Vocke, G., E.O. Heady, W. Boggess, and H. Stockdale. Economic Impacts on Agriculture from Pesticide, Fertilizer, Soil Loss and Animal Waste Regulating Policy, CARD Report 73, Iowa State University, 1977.

Russell G. Thompson, John C. Stone,
Sankar Muthukrishnan, Srikant Raghavan

8. Input-Output Modeling from the Bottom Up Rather Than from the Top Down

Introduction

Leading econometricians have attempted to join the Keynesian and Walrasian schools of economic thought by including micro-economic structures in economy-wide macroeconomic models. Hudson and Jorgenson (7) and Hoffman and Jorgenson (6) have sought to bridge this gap for energy policy and economic growth issues by including simple process models directly in a national econometric model of the U.S. economy. Griffin (4) fitted regression equations to pseudo data (from process models) and used the regression equations to modify input-output coefficients. Klein (10) in his Presidential address to the ninetieth meeting of the American Economics Association said:

> A principle feature of such combined systems is that they are not based on restrictive assumptions of the fixed coefficient input-output model, but are generalized to allow the coefficients of production to vary, according to the variation of relative prices. (10, p. 4)

The focal point of the Keynesian and Walrasian schools of thought is the Leontief transactions table. It is here that the microeconomic determinants of supply, demand, and price must sum up to fulfill the fundamental balance equations between total inputs and total outputs. If the balance equations are not fulfilled, then prices must adjust to achieve the necessary balance between final demands, value-added, and interindustry transactions.

This study examines the same concepts as those of econometric modelers of the economy. Instead of estimating all transactions values by disaggregating relationships from an econometric model of the whole economy, key transactions values are here estimated by aggregating relationships from a detailed equilibrium model of the energy sector, which is an important part of the national economy. The transactions table depicts the energy sector as a microcosm economy in the following way: All purchases of inputs by this economy from sectors outside the microcosm are regarded as imports; all sales of outputs from this economy to sectors outside the microcosm are regarded as exports.

This bottom-up approach to estimation of the transactions table allows us to apply the fundamentals of econometrics, as developed by Marshak (12). These fundamentals allow the coefficients of production to change in response to the technical effects of both technology-based policies and economics-based policies. For example, best available technology mandates for pollution control require the use of specific new processes in production, whereas legislated restrictions on fuel use in boilers prohibit the use of existing processes (or investment in new versions of these processes). This type of change in the technical coefficients of production modifies both the domain and the range of the production function and in turn both the domains and ranges of the derived factor demand and endproduct supply functions. In linear programming terms, these technology-based policies add structural constraints to the allocation problem and additional value terms to the valuation problem. The additional value terms increase the value-added of the sector being affected; they further increase the level of prices in the economy. Indirectly, these higher price levels affect the technical coefficients in the input-output table. It is this important indirect effect that economy-wide modelers have not yet incorporated in their econometric analyses. To overlook this indirect price effect is to ignore one of the primary economic problems of our age.

The goal of this study is to show how bottom-up modeling may be used to capture significant elements of the indirect price effect, as needed to evaluate the macroeconomic consequences of technical change. Specifically, the objectives for this study are threefold:

- To show how the equilibrium solution from a detailed structural model may be used to develop a transactions table for the industries in the energy sector;

- To show how economic process models of the central energy conversion industries may be combined with energy resource supply models and endproduct demand models to represent a competitive equilibrium model for the fossil energy sector; and

- To show how (a) three different energy policies affect the structure of the equilibrium model, (b) these different structures affect the price/quantity solutions to the equilibrium model, and (c) the different solutions affect the transactions tables and resulting input requirements.

The organization of the paper is as follows. First, a brief review of the modern approach to general competitive equilibrium gives a clear frame of reference for the Leontief balance equation. This economy-wide perspective shows how the economic process models provide the technical change information required to modify the technical coefficients in an input-output transactions table. It also provides a backdrop for describing how the Leontief balance equation may be developed for an energy microcosm economy.

Second, methods of analysis are described for integrating economic process models of conversion industries, resource supply models, and energy endproduct demand models into a competitive equilibrium model for the energy sector. Procedures are outlined for extracting relevant input-output information from an equilibrium solution to this model; relevant information includes interindustry transactions (within the microcosm) and the total value of inputs and outputs for each industry and the microcosm as a whole.

Third, an application of this methodology is presented. Three different policy structures provide the basis for three price-quantity equilibrium solutions. Price-quantity information from these solutions is used to derive transactions tables for each structure, and differences in these resulting tables illustrate the importance of structural determination in an interface of micro- and macroeconomic models.

Leontief Balance Equation and Fossil Energy Microcosm

Using Intriligator's presentation (8) of the modern input-output/linear programming approach to general competitive equilibrium, the Leontief balance equation is one part of the solution to the equilibrium problem. This balance equation may be stated in tabular form to show how interindustry transactions, value-added, and final demands must aggregate into total inputs and outputs for any competitive equilibrium solution. Use of this tabular form demonstrates clearly how the competitive equilibrium problem for a microcosm energy economy fits into a general equilibrium framework.

Review of Input-Output

This demonstration may be portrayed for a two-sector energy and nonenergy economy in the following way (with all values in dollar terms). Let x_1 = the energy output; x_2 = the nonenergy output; p = the vector of unit prices of energy, nonenergy, labor, and capital; $c_1(p)$ = the aggregate demand function for energy; $c_2(p)$ = the aggregate demand function for nonenergy; v_1 = the value-added

Table 1

Illustrative Input-Output Table for the Economy+

	Energy	Nonenergy	Demands	Outputs
Energy	$a_{11} x_1^*$	$a_{12} x_2^*$	c_1^*	x_1^*
Nonenergy	$a_{21} x_1^*$	$a_{22} x_2^*$	c_2^*	x_2^*
Value-Added	v_1^*	v_2^*	N.P. =N.I.	
Total Inputs	x_1^*	x_2^*		

+ $v_1 = b_{11} x_1 + b_{21} x_1$ and $v_2 = b_{12} x_2 + b_{22} x_2$

by the energy sector; and v_2 = the value-added by the nonenergy sector. Further, let the technical coefficients of industrial production a_{ij} (i, j = 1, 2) establish the proportionality relationships among interindustry outputs and inputs:

$$x_{ij} = a_{ij} x_j \qquad i, j = 1, 2.$$

For example, the input of industry sector j purchased from industry sector i is a fixed proportion a_{ij} of the jth sector's output x_j. Similar proportionality relationships may be established between the level of primary input use in the economy and its use in each industry by the technical coefficients of primary input use b_{ij}(i, j, = 1, 2).

With stars representing equilibrium values, the transactions table for the economy may be stated as shown in Table 1. Table 1 shows the following relationships of input-output analysis:

- Total input purchases by each sector equal the sum of purchases by that sector from itself and from all other sectors plus the purchases of services by that sector from holders of capital and labor.

- Total outputs of each sector equal the sum of sales by that sector to itself, to other sectors, and to final demand, where final demand includes consumption, investment, exports, and government demand.

- Total outputs of each sector equal total inputs of that sector.

- National production equals the sum of the values of final demands.

- National income equals the sum of the value-added by each sector in the purchase of primary input services.

- National income equals national product.

The equalities across each row of the transactions table give the Leontief equations for the energy and nonenergy sectors:

$$a_{11} x_1^* + a_{12} x_2^* + c_1^* = x_1^*$$

$$a_{21} x_1^* + a_{22} x_2^* + c_2^* = x_2^*.$$

These Leontief balance equations hold exactly at a competitive equilibrium price/quantity solution for the economy. They assume a constant input-output structure for interindustry transactions (because of the proportionality assumption).

Variable Input-Output Coefficients

In the simplest case, the a_{ij} coefficients are treated as constants. A more sophisticated analysis recognizes that the a_{ij} are variables determined by underlying producer optimizations in the industrial sectors. Specifying a functional relationship between the a_{ij} and equilibrium prices allows modeling of the structural effects of a limited class of government policies; these may be referred to as economically based policies. Such policies directly alter the prices or cost coefficients in the producer optimization problems and may thus, with proper microeconomic modeling, be mapped into a revised set of a_{ij} coefficients. This price dependence for the a_{ij} has been incorporated in certain top-down modeling efforts to capture the effects of economically based policies.

There is, however, a more fundamental source of variation in the a_{ij} which to date has not received much attention in applied econometric modeling. As recognized by Marshak (12), the functional dependence of the a_{ij} on prices is itself dependent on the underlying technical structure of production. A wide class of government policies directly affect this technical structure by requiring or prohibiting certain kinds of industrial operations. These technology-based policies directly alter the set of options open to the producer and thereby change the pattern of price dependence. Both the domain and the range of the production function may be altered as well as its specific form and properties, although the induced alteration may not be continuous. The transformed production structure creates a different set of equilibrium prices with consequent indirect effects on all input-output coefficients in the system.

To capture the structural effects of both classes of policies, the solution procedure for a competitive equilibrium must be augmented to include detailed models of each sector capable of representing the structural transformations of technology-based policies as well as the price-responsiveness of production. This augmentation effectively generates new data for estimating the a_{ij}. Such data are used in this study to estimate the structural coefficients for the energy sector. This structural coefficient estimation and the equilibrium price/quantity determination are simultaneously accomplished for all of the sectors in the energy microcosm.

Illustration

This simultaneous determination may be illustrated by specifying a linear programming problem for the energy sector:

Primal L. P. Problem

$$\min_{y} \underset{(1\times n)(n\times 1)}{cy} \qquad \text{Cost of production}$$

s.t. (2.1)

$$\underset{(m\times n)(n\times 1)}{Qy} - \underset{(m\times n)(n\times 1)}{My} \geq \underset{(m\times 1)}{d^T(p^*)} \qquad \text{Demand requirements for energy endproducts}$$

$$\underset{(s\times n)(n\times 1)}{Ny} \leq \underset{(s\times 1)}{r^T(p^*)} \qquad \text{Availabilities of energy resource inputs}$$

$$y \geq 0$$

The primal allocation problem is to minimize the costs of producing the demand requirements for outputs subject to the limited availabilities of energy resource inputs. Both demand requirements and resource availabilities are determined for the equilibrium prices, p^*. The unit cost vector, c, includes the capital (c_K), labor (c_L), and other (c_O) input cost components for each activity; it is assumed that these costs are based on economy-wide equilibrium prices. All terms in Q, M, and N are nonnegative parameters.

The dual L. P. problem has the following form:

Dual L. P. Problem

$$\max_{u,\, w} \; d(p^*)u - r(p^*)w$$

s.t. (2.2)

$$Q^Tu - M^Tu - N^Tw \leq c^T \qquad \text{No profit constraint}$$

$$u \geq 0, \; w \geq 0,$$

where u is the vector of marginal output costs and w is the vector of marginal input values.

Application of the fundamental L. P. theorems solves the primal and dual problems and the relationships between the two solutions; see Duffin et al. (2, Thms. 1, 2, 3,). In practice, iterative solutions to the competitive equilibrium problem may be necessary to establish consistency between prices in the vector p*, which are endogenous to the energy sector, and shadow prices in the dual vectors u and w. That L. P. solution gives the following solution for the energy sector terms in Table 1:

- Total energy output $x_1^* = (Qy^*)^T u^* = d(p^*)u^* + (My^*)^T u^*$
- Total energy input

  $$x_1^* = r(p^*)\, w^* + (My^*)^T u^* + c_L y^* + c_K y^* + c_O y^*$$
- Value-added $v_1^* = c_L y^* + c_K y^*$
- $a_{11} = [r(p^*)\, w^* + (My^*)^T u^*] / x_1^*$ (2.3)
- $a_{21} = c_O y^* / x_1^*$
- $b_{11} = c_L y^* / x_1^*$
- $b_{21} = c_K y^* / x_1^*$

Clearly, all of the terms in the energy sector column of Table 1 are determined (as functions of the equilibrium prices) from an equilibrium solution of the primal and dual L. P. problems. This is the variable coefficient problem referred to by Klein (10).

Inclusion of a set of noneconomic constraints of the form:

$$Ay \leq b^T \qquad (2.4)$$

as an additional set of row constraints in the primal L. P. problem allows the variable coefficient input-output problem to include technology-based policy effects. This extension may include pollution control processes internalized by industry in response to government regulations, as posited by Leontief (11, Section IX). It may also include boiler fuel restrictions and other direct technology controls by government regulation.

The modified L. P. problems have the following forms, where z is the marginal value vector for the noneconomic constraints:

Modified Primal	Modified Dual
$\min_{y} c\,y$	$\max_{u,w,z} d(p^*)u - r(p^*)w - bz$
s.t.	s.t.
$Qy - My \geq d^T(p^*)$	$Q^Tu - M^Tu - N^Tw - A^Tz \leq c^T$
$Ny \leq r^T(p^*)$	$u \geq 0,\ w \geq 0,\ z \geq 0$
$Ay \leq b^T$	
$y \geq 0$	

(2.5)

Costs of each production process in the energy sector are increased by the additional value-added component contributed by noneconomic constraints, A^Tz^*. The total value-added of the energy sector is increased by the total value imputed to the noneconomic constraints, bz^*. This additional value-added must be included in the calculation of total input.

Noneconomic constraints (if binding) change the values of u^* and w^* in the modified dual (and the value of y^* in the modified primal). These direct changes, in addition to the new dual variables z^*, modify the price vector p^*, which subsumes the cost vector c. These modified prices change the demand requirements and resource availabilities in the energy sector in particular. Thus, all of the terms in the energy sector column in (2.3) are subject to change.

In addition, noneconomic constraints for the energy sector may indirectly affect, by way of price adjustments, the final demands for the nonenergy sector and the technical coefficients in the nonenergy column. Such adjustments in the nonenergy sector are not considered in this study.

It is noteworthy that constraints reflecting the fixity and heterogeneity of past investments in production capacity have the same linear form as that specified for the noneconomic constraints. They accordingly have a similar effect on equilibrium solutions and implied transactions. These existing capacity restrictions have also been ignored in the basically flow-oriented methodologies of neoclassical economic analysis. Ignoring these constraints has thus resulted in another source of bias in economic analyses of structural change.

Methods of Analysis

Detailed documentation of the conversion industry, resource supply, and final demand models incorporated in the competitive equilibrium model for the fossil energy sector was initially provided in three monographs by Thompson *et al.* (13, 14, 15). Kim and Thompson (9) documented the supply model currently used for new onshore crude oil and natural gas supplies in the lower 48 states. This documentation was updated for the industry models, the coal supply models, and the equilibrium computational procedure in an analysis of President Carter's National Energy Plan by Thompson *et al.* (16).

Equilibrium Model for Energy Microcosm

The equilibrium model for the energy sector is built around an Integrated Industry Model for the electric power, petroleum refining, and basic chemicals industries. The Integrated Industry Model is a large-scale, linear programming production model in which the objective is to minimize production costs subject to constraints. Constraints include output requirements, limitations on waste discharges to the air and water, materials balances, production capacities, and resource availabilities. Mathematically, the economic allocation model may be defined as follows:

$$\text{Minimize} \qquad C^T X$$

subject to the restrictions:

$$
\begin{array}{lll}
\text{(a)} & A_e X \leq B_e & \text{(technical, environmental)} \\
\text{(b)} & A_k X \leq K & \text{(production capacity)} \\
\text{(c)} & A_q X \leq Q & \text{(availability of nonenergy inputs)} \\
\text{(d)} & A_s X \leq S & \text{(availability of energy inputs)} \\
\text{(e)} & A_d X \geq D & \text{(requirements for outputs)} \\
 & X \geq O, &
\end{array}
\qquad (3.1)
$$

where C = vector of unit costs of production processes and resources,

A = matrix of process input-output coefficients, and

X = vector of process activity levels.

The dual of this allocation program results in the following valuation model:

$$\text{Maximize} \quad -B_e^T u - K^T v - Q^T w - S^T \pi_s + D^T \pi_d$$

subject to the restrictions: (3.2)

$$(f)\ -A_e^T u - A_k^T v - A_q^T w - A_s^T \pi_s + A_d^T \pi_d \leq C$$

$$u, v, w, \pi_s, \pi_d, \geq 0$$

The primal and dual problems determine the (shadow) prices, π_s and π_d, as mappings of S and D. Specifically, they determine marginal value of energy inputs,

$$\pi_s = F(S, D; Be, K, Q), \tag{3.3}$$

and the marginal cost of energy products,

$$\pi_d = G(D, S; Be, K, Q), \tag{3.4}$$

where π_s = the computed price (marginal value) of an additional unit of energy resources and;

π_d = the computed price (marginal cost) of an additional unit of endproduct requirements.

Both mappings are further dependent on the structure of production as represented by the A matrices.

Additionally, exogenous estimates of energy resource supplies are specified as functions of the prices of oil, natural gas and coal; and exogenous estimates of final demand for energy products are specified as functions of endproduct prices. These can be represented by supply functions for crude oil, natural gas, and coal,

$$S = H(P_s; E), \tag{3.5}$$

and demand functions for important energy endproducts,

$$D = T(P_d; O), \tag{3.6}$$

where S = supply of oil, natural gas, coal;

P_s = prices of oil, natural gas, coal;

E = other factors affecting supply (e.g., technology, rates of return);

D = demand for energy;

P_d = prices of energy products; and

O = other factors affecting demand (e.g., population, income).

For a competitive equilibrium solution, a point must be found on the supply and demand curves which is also a solution to the linear program; furthermore, this solution must be such that:

(1) Supply and demand prices are equal, i.e.,

$$P_s = \pi_s \text{ and}$$

$$P_d = \pi_d;$$

(2) There is no excess demand for energy resources or endproducts; and

(3) Excess supply of a resource or endproduct, if it exists, is accompanied by a zero price for that resource or endproduct.

Condition (1) means that the shadow prices from the linear programming model and the input prices (to the supply and demand models) for energy resources and endproducts must be equal.

It must be made clear that the solution so defined is an equilibrium for the energy sector microcosm only, since the only variables considered are the supply and demand prices for energy. Other factors affecting the supply and demand for energy are assumed to be determined outside the model. Consistency with a general equilibrium is assumed.

In the case of a regulated supply market, conditions (1) and (2) break down to the extent that excess demand will exist in any market for which the supply price is fixed lower than the market-clearing price. In this situation, "equilibrium" must be understood as a stationary point at which: (i) supply and demand prices are equal in all non-regulated markets; (ii) supply prices in the regulated markets are equal to a policy-specified constant, and (iii) demand prices in the regulated markets are those demand prices corresponding to the available quantity. These demand prices are not directly paid but serve in the analysis to allocate the available supply and to shift marginal demanders to alternative fuels.

Integrated Industry Model

The Integrated Industry Model (IIM) organizes data on selected production and support processes in several energy- and water-intensive industries. Products included in the model (as used in this analysis) are identified in Table 2. This table classifies the products according to the functional groupings employed in the IIM; as an additional frame of reference, the Standard Industrial Classifications for which each grouping is a subset are also identified. The choice of products for inclusion in the model (over a number of years of development) is based on three principal criteria: (1) volume of output, (2) completion of a chain of chemical processing, and (3) particular importance in terms of energy and water use.

Production of the indicated product requires consumption of a number of primary resources: crude oil, natural gas, gas liquids, coal, and water. The IIM is interfaced with supply models for the energy resources as part of the equilibrium system described previously. Water can be constrained or priced as desired, but this is not undertaken in the context of the present analysis. Other primary and intermediate inputs are required by the model, but only their costs are considered.

Each of the functional groups in the IIM (to be referred to as industries) are structured in a similar manner. Each is composed of interrelated systems for production, process energy use, water use, and waste management (solid, liquid, and gaseous). Modeling important substitution possibilities within each system allows cost-optimal design and operation of the overall complex of systems. Industry competition for common resources establishes an important interdependence of all modeled operations.

Optimal substitutions within and between industries and systems are restricted by both limited availability of resources and various classes of technical constraints. Constraints include: (a) limits on the availability of existing production capacity, (b) logical and materials balance conditions, and (c) technology-based policy constraints.

Exhaustion of existing capacity requires substitution of an alternative process or investment in capacity expansion. The supply of capital is not constrained in this analysis (completely price elastic), and capital charges include a normal rate of return on investment.

Logical and materials balance conditions are specified for all intermediate product, energy, water, and waste flows. These conditions are required in any linear programming model to prevent artificial augmentation and diminution of desirable and undesirable flows, respectively. Such conditions are important to ensure an accurate and consistent mapping of resources into products.

Properly specified structural constraints are essential to modeling technology-based policies. Depending on the application, the IIM can be structured to incorporate limits on waste discharges (total or per unit of process activity), prohibitions on the use of

Table 2

PRODUCTS INCLUDED IN INTEGRATED INDUSTRY MODEL
(By Functional Group and with Relevant Standard Industrial Classification)

Product	SIC
Electricity	4911
Natural gas	4923
Petroleum refining products	2911
Asphalt	
Butane (mixed, normal, iso*)	
Distillate fuel oil	
Gasoline (premium and regular)	
Grease	
Kerosene (including jet grade)	
Liquified petroleum/refinery gas	
Lubricating oil	
Naphtha (including jet grade)	
Petroleum coke	
Residual fuel oil	
Wax	
Organic chemicals	2896,2865
Butadiene	
Butylene (mixed, normal, iso)	
Ethylene oxide	
Ethylene glycol	
Methanol	
Propylene	
Benzene	
Cyclohexane	
Ethylbenzene	
Phthalic anhydride	
Styrene	
Toluene	
Xylene (mixed, ortho, meta, para)	

*Arguably in SIC 2819

Table 2 (Concluded)

Plastics and polyesters	2821,2824
Polyethylene—high density	
Polyethylene—low density	
Polymer wax	
Polypropylene—copolymer	
Polypropylene—homopolymer	
Polystyrene—expandable	
Polystyrene—general purpose	
Polystyrene—high impact	
Polyvinyl chloride	
Vinyl chloride monomer	
Adipic acid	
Dimethyl terephthalate	
Nylon (66)	
Polyester fiber	
Terephthalic acid	
Alkalies and chlorine	2812
Caustic soda	
Chlorine	
Lime	
Soda ash	
Nitrogenous fertilizers	2873
Ammonia	
Ammonium nitrate	
Nitric acid	
Urea	
Inorganic chemicals	2819
Alum	
Sulfuric acid	

particular processes, or requirements that certain processes be used. The particular policy structures imposed for the present analysis are discussed in a later section.

The various industries in the IIM are linked together by rows representing common resources and intermediate products that are transferred from one industry to another. (Electricity generated internally by a non-utility industry may or may not be transferred to another industry, at user discretion.) Flows of water, waste products, steam, shaft energy, and the like are not transferred between industries. Model structure specifies that all transfers take place at marginal cost (with the possible inclusion of a transportation or other add-on charge). This transfer structure is central to the mapping of input costs into output costs and is the basis for calculating interindustry transactions from a model solution.

The IIM is by nature a static optimization model. The time frame for an analysis is essentially specified via the product demands, resource availabilities and prices, existing capacities, and those capacity expansion options that are deemed reasonable and feasible. The short- or long-run nature of the model is thus essentially at user discretion. Application to long-term time horizons (say, beyond 2000) is limited by the relevance of the modeled technology options and would require appropriate extension of the modeling base to include speculative technologies.

Energy Resource Supply Models

Price-sensitive supplies for crude oil, natural gas, and coal are derived from three modeling bases: the Kim and Thompson oil/gas supply model (9), the oil and gas supply model of the Federal Energy Administration, and the coal model of FEA as adapted at the University of Houston (16, Appendix).

For the present study, coal supplies are based on data obtained by personal communication with FEA (later published (3)). These data consist of step function supply curves for a number of different coal types in a number of different regions. Each step function relates minimum acceptable selling prices to available quantities of coal at a given level of extraction difficulty. These step functions are re-aggregated along type and regional lines for consistency with the structure of the IIM. The resulting piecewise-specified supply curves are incorporated directly in the linear programming model.

The FEA oil and gas supply model is used for specification of supplies from old reserves, offshore fields, and Alaska. As modeled by FEA, old field supplies are not price-sensitive and offshore supplies are lease-constrained at the level of prices considered in this study. These supply estimates are obtained from information reported in the 1976 National Energy Outlook.

The Kim and Thompson model is used to obtain price-sensitive supplies of oil and gas from new fields in the lower 48 states. This model treats oil and gas as joint products, and determines a time stream of production based on a producers' objective of maximizing a discounted stream of profits. This maximization entails determination of an optimal level of exploratory drilling and of an optimal depletion rate for production from discovered reserves. New discoveries are determined by drilling levels and an exponentially declining finding rate based on the historical record. Repeated solutions of the model for different oil and gas prices defines a joint supply schedule for any desired year.

Endproduct Demand Model

Price-sensitive demands for energy endproducts are derived from an adaptation of the (PIES) final demand model employed by the Federal Energy Administration in the 1976 National Energy Outlook. Products covered in this adaptation are: electricity, gasoline, distillate fuel oil, residual fuel oil, natural gas, LPG, kerosene, jet fuel, and coal. The model is a constant elasticity approximation of an econometrically estimated, dynamic demand model. The approximation consists of a set of base demands and base prices and a matrix of own- and cross-price elasticities; these values correspond to a particular solution of the PIES equilibrium system.

For the present study, sufficient information was available only for the $13 Reference Scenario of the 1976 National Energy Outlook. Regionally specified demands are aggregated to two regions for consistency with the two-region structure of the electric utility sector in the IIM. Demands for coal and electricity are specified regionally in the IIM, while demands for other products are aggregated to a national total. That portion of base demand expected to be accounted for by the chemicals production components of the IIM is subtracted to avoid double-counting. Base prices for the two regions are derived as quantity-weighted averages of regionally specified base prices in the FEA documentation. Only one elasticity matrix was available, however, reportedly a weighted average of regional elasticity matrices. The resulting elasticities and base prices and demands specify a vector-valued demand function for use in the competitive equilibrium system.

Constant demands for the chemical and other products in the IIM are derived by extrapolating 1975 production levels reported in the literature. This extrapolation is based on an assumed three percent real economic growth rate (1.2 percent population, 1.8 percent per capita income) and income elasticities reported by Almon (1).

Table 3

Transactions Table for Energy Microcosm

	Coal Supply	Petroleum Supply	Petroleum Refining	Electric Utilities	Modeled Chemicals	Final Demand	Total Outputs
Coal Supply	*	*	a	a	a	d	TO
Petroleum Supply	*	*	a	a	a	d	TO
Petroleum Refining	*	*	a	a	a	d	TO
Electric Utilities	*	*	a	a	a	d	TO
Modeled Chemicals	*	*	a	a	a	d	TO
Payments Sector	*	*	a	a	a		
Total Inputs	TI	TI	TI	TI	TI		

Input-Output Framework for Energy Microcosm

An input-output tabulation for the competitive equilibrium system of the energy microcosm is developed as a straightforward extension of the two-sector example presented earlier. The energy sector is partitioned into four key subsectors: coal supply, petroleum supply, petroleum refining, and electric utilities. (Petroleum supply includes crude oil, natural gas, and gas liquids from both domestic and imported sources.) The nonenergy sector is represented in two ways. First, a specific industrial sector is defined to encompass transactions in the production of chemical products included in the Integrated Industry Model (IIM). Second, the transactions of all other industrial sectors are treated analogously to imports and exports. By this analogy, purchases of energy by these sectors are included in the final demand column, while purchases from these sectors (by the explicitly defined energy and chemicals sectors) are included in the payments sector. Other components of the final demand column are household and commercial purchases, government purchases, and exports to other countries. Other components of the payments sector are value-added, nonpetroleum imports from abroad, taxes, depreciation, and the imputed cost of noneconomic constraints (both technology-based policy considerations and existing production capacities).

In accordance with this accounting scheme, the input-output structure of the energy microcosm is represented as a 5x5 transactions matrix plus the additional rows and columns for final demand, the payments sector, and total inputs and outputs; see Table 3. All values in this table except those indicated by "*" are computed from the price/quantity solution to the equilibrium system by procedures analogous to those discussed earlier. Total outputs of the coal and petroleum supply sectors are also known from the equilibrium solution; after equating total inputs to total outputs, the transactions values for these sectors (*) are determined by applying the proportions reported by the U. S. Department of Commerce in the Input-Output Structure of the U.S. Economy: 1967 (17). This procedure assumes that the structure of the coal and petroleum supply sectors is not significantly affected by the policy and price variations under study.

Application of Methodology

As an illustration of the methodology presented in this paper, the relevant input-output structures are derived for three distinct solutions of the competitive equilibrium system for the energy microcosm. Identifying specifications for each solution are outlined first, followed by a tabulation of key price-quantity values from the solutions and the input-output structures derived from the three solutions.

Specifications and Assumptions

The equilibrium system and solutions employed in this illustration are extracted from an analysis of President Carter's National Energy Plan by Thompson et al. (16). Detailed specifications and assumptions for the solutions are available in the referenced report. In brief, the solutions reflect equilibrium solutions (static) for 1985 under three alternative policy scenarios:

(1) a business-as-usual scenario (BAU) modeling continued regulation of domestic oil and gas prices;

(2) a complete deregulation scenario (CPD) under which new domestic oil and gas prices are deregulated; and

(3) a modified price regulation scenario (MPR) capturing important features of the National Energy Plan.

Key structural specifications on the resource supply side are the price restrictions for the regulated cases. These (economically based) policy specifications are (in 1975 dollars):

	BAU	MPR	CPD
New Natural Gas ($/Mcf)	1.42	1.61	Market-Clearing
Crude Oil from pre '74 Reserves (L. Tier) ($/bbl)	4.84	4.84	Market-Clearing
Crude Oil from '74-77 Reserves (U. Tier) ($/bbl)	10.40	10.40	Market-Clearing
Crude Oil from New Reserves (after '77) ($/bbl)	10.40	13.78	Market-Clearing

Key structural specifications on the demand side are the base demands, base prices and elasticities for the final demand model, and a set of excise taxes imposed under the MPR scenario. These taxes (in 1975 dollars) are represented in both the final demand model and the Integrated Industry Model:

	Transportation ($/bbl)	Industry ($/MMBtu)	Electric Utilities ($/MMBtu)
Oil Products	-	.46	.20
Natural Gas	-	1.54	1.24
Gasoline	14.70	-	-

The taxes imposed on industry are applied only to petroleum hydrocarbons used as fuels. Chemical feedstocks are exempt from taxation, as are all synthetic fuels derived from coal.

A number of important technology-based policy constraints are incorporated directly into the structure of the Integrated Industry Model:

- Use of oil products and natural gas in all new baseload electric power plants and new industrial steam boilers is prohibited in all three cases.
- Cogeneration of electricity by the process industries for sale into the grid of electric utilities is allowed in all three cases.
- Treatment by use of the best available technology is required for all major waste-water effluents.
- The sulfur content of oil sold to residential/commercial (and non-modeled industrial) sectors must meet new source air emission standards. The sulfur content of coal sold to these sectors must be equivalent to no more than 1.25 times the new source sulfur oxide standards.
- Air emissions from new plants must meet new source standards as promulgated by EPA. In the Modified Price Regulation scenario, stack gas scrubbers are required on all new fossil-fueled electric power plants and new coal-fired industrial steam boilers. In all cases, old plants must burn oil having 1% sulfur or less, and old coal-fired plants must not emit more than one and one-half times as much sulfur oxides as allowed by new source standards.

Formulating and applying the supply, demand, and industry models in accordance with the above specifications defines three

Table 4

Estimated 1985 Energy Supplies and Marginal Values*

Supplies	BAU Annual Qty	BAU Marginal Value	MPR Annual Qty	MPR Marginal Value	CPD Annual Qty	CPD Marginal Value
Crude Oil	Billion BBL	$/BBL	Billion BBL	$/BBL	Billion BBL	$/BBL
Domestic	3.97	13.75	4.10	13.84	4.04	13.78
Imported	2.42	13.00	1.94	13.00	2.29	13.00
Imported Resid	0.19	13.00	0.29	13.00	0.21	13.00
Nat. Gas liquids	0.55	**	0.56	**	0.63	**
Natural gas (dry marketed)	Trillion cu ft	$/Mcf	Trillion cu ft	$/Mcf	Trillion cu ft	$/Mcf
Domestic	17.40	2.35+	17.75	2.34+	19.38	2.29+
Imported	1.84	2.50	0.01	2.50	0.0	
Coal	Million tons	$/ton	Million tons	$/ton	Million tons	$/ton
Eastern High Sulfur	54	10.80	115	14.05	54	10.80
Eastern Low Sulfur	310	17.50	309	17.50	306	17.50
Midwestern High Sulfur	133	10.38	155	14.91	133	10.38
Western Low Sulfur	452	8.90	230	7.30	448	8.90

* 1975 dollars.
** Priced in fixed proportion to crude oil.
\+ Before processing.

structurally distinct equilibrium systems for the energy microcosm. Solutions to these systems are obtained by iterative procedures described in the referenced report. The input-output structures implied by these equilibria are derived from the linear programming solutions and are presented below.

Equilibrium Results for Energy Microcosm

Important prices and quantities from the three equilibrium solutions are presented in Tables 4 and 5; the former focuses on energy resource supplies, and the latter focuses on selected energy endproducts. It should be noted that marginal values and costs are tabulated, which by assumption are equivalent to prices (before distribution) in competitive markets. The marginal values for oil and gas are not equal to input prices in the regulated cases (BAU, MPR).

Given the price-limiting effect of perfectly elastic import supplies, it is to be expected that the marginal values of oil and gas would be similar in all cases. (The difference in value between domestic and imported oil reflects a quality difference.) An increase in domestic quantity is observed in response to higher allowed input prices. The significantly higher price for gas under CPD actually decreases the quantity of oil from the joint supply function.

The impact of the scrubber requirements and excise taxes under MPR is readily apparent. As can be seen in Table 4, the uniform scrubber requirements induce a significant shift from Western low sulfur coal to higher sulfur coals from the East and Midwest; this shift is accompanied by equilibrium price adjustments. It is interesting, however, that the total expenditure on coal remains fairly constant across the three cases (11.42, 11.02, and 11.30 billion 1975 dollars under BAU, MPR, and CPD, respectively). This is indicative of the kinds of structural effects which may be obscured by too coarse an aggregation in the input-output structure. (With this effect noted here, the input-output tables below aggregate the coal supplies for convenience.)

The impact of the excise taxes under MPR is readily apparent in Table 5 as a depression for demand for the taxed commodities and for electricity (for which the marginal cost increases due to both the excise taxes on fuels and the uniform scrubber requirements). These demand effects are in turn translated into lower equilibrium quantities for crude oil, natural gas, and coal; see Table 4.

Transactions Values for Input-Output Table

Marginal values of energy factor inputs, marginal costs of energy endproduct demands, and the corresponding equilibrium

Table 5

Estimated 1985 Final Demands and Marginal Costs*

Demands	BAU		MPR		CPD	
	Annual Qty	Marginal Value	Annual Qty	Marginal Value	Annual Qty	Marginal Value
Electricity (billion kWh)	3240.7	$.023/kWh	3079.7	$.027/kWh	3239.6	$ .023/kWh
Natural gas (trillion cu ft)	14.86	$2.50/MCF	13.74	$2.50/MCF	15.0	$2.44/MCF
Coal (Quads)	3.14	$1.17/MMBtu	3.47	$1.14/MMBtu	3.13	$1.17/MMBtu
Resid. (low sulfur) (Billion Barrels)	0.52	$14.85/BBL	0.51	$15.06/BBL	0.52	$14.89/BBL
Distillate (low sulfur) (Billion Barrels)	1.35	$15.31/BBL	1.36	$15.63/BBL	1.35	$15.34/BBL
Gasoline (Billion Barrels)	2.79	$13.3/BBL	2.38	$13.62/BBL	2.79	$13.4/BBL

* 1975 dollars

quantities are the primary inputs used to compute the transactions values for the CPD Case. This is illustrated in Table 6 for input purchases of petroleum supplies by the petroleum refining sector. The petroleum refining sector in the CPD Case purchases 2.5 billion barrels of imported crude oil at a price of $13 per barrel for a transactions value of $32.53 billion. Also, it purchases 4.04 billion barrels of domestic crude oil at a price of $13.79 per barrel for a transactions value of $55.71. In addition, the petroleum refining sector purchases 1.77 trillion cubic feet of natural gas at $1.74 per Mcf. This price of $1.74 represents a weighted average price for new natural gas at $2.29 per Mcf and old natural gas at $.80 per Mcf. The transactions value of the natural gas purchases by the petroleum refining sector is $3.08 billion. Summing the input purchases of imported crude oil, domestic crude oil, and natural gas gives a transactions value of $91.32 billion in 1975 dollars or $66,785 million in 1967 dollars. This transactions value of $66,785 million is shown to be purchased by the petroleum refining sector from the petroleum supply sector in Table 9. Similar procedures are used to compute the transactions values for petroleum supplies in the MPR and BAU Cases. A lower average input price exists for natural gas in the MPR Case than in the CPD Case because new natural gas is regulated at $1.75 per Mcf. The average price for natural gas is even lower in the BAU Case because of the lower regulated price for new natural gas ($1.42 per Mcf). Similarly, continued regulation of domestic oil prices under BAU (at $4.84 per barrel for lower tier oil and $10.40 per barrel for upper tier oil and oil from new reserves) is the reason for an average oil input price of $8.40 per barrel. The resulting transactions values of $64,384 million in the MPR Case and $50,817 million in the BAU Case (in 1967 dollars) are shown to be purchased by the petroleum refining sector from the petroleum supply sector in Tables 8 and 7, respectively.

Similar procedures are used to compute the transactions values of purchases by the petroleum refining sector of its own output as well as purchases from the coal supply, electric utilities, and modeled chemicals sectors. The transaction value of the petroleum refining sector in the payments sector represents payments for labor, capital, non-petroleum raw materials, and land (as reflected in the cost coefficient of the objective function) plus the imputed cost of non-economic constraints. This transaction value must be calculated as the residual between total outputs and the total of inputs from all sectors other than the payments sector (which are directly calculated as described above). Both total outputs and final demand are directly calculated from the model solution by the matrix operations indicated earlier. This set of procedures is repeated to compute all of the interindustry transactions, payment values, total inputs, and total outputs for the electric utilities and modeled chemicals sectors.

Table 6

Inputs Purchases of Petroleum Supplies by the Petroleum Refining Sector of BAU, MPR, and CPD Modeling Cases

Input Purchases by Sector		Modeling Cases								
		Business As Usual (BAU)			Modified Price Regulation (MPR)			Complete Price Deregulation (CPD)		
Petroleum Mining	Units	Price ($/unit)	Amount (bils of units)	Value (bils of $)	Price ($/unit)	Amount (bils of units)	Value (bils of $)	Price ($/unit)	Amount (bils of units)	Value (bils of $)
Imported Crude Oil	(BBL)	$13.00	2.61	33.956	$13.00	2.23	29.012	$13.00	2.50	32.53
Domestic Crude Oil	(BBL)	$ 8.40	3.97	33.35	$13.84	4.10	56.744	$13.78	4.04	55.71
Natural Gas	(SMCF)	$ 1.17	1.86	2.18	$ 1.29	1.77	2.28	$ 1.74	1.77	3.08
Total Input	(bils of 1975 $)		69.486			88.036			91.32	
Transfer Value	(mils of 1967 $)		50,817			64,384			66,785	

Only total outputs of the coal supply and petroleum supply sectors are available from the model. After equating total outputs to total inputs, input purchases are distributed across the coal supply, petroleum supply, petroleum refining, electric utilities, modeled chemicals, and payments sectors by applying the proportions in the Input-Output Structure of the U.S. Economy: 1967 (17).

Significant differences clearly exist between the transactions values in Tables 7, 8, and 9. For example, input purchases by the petroleum refining sector from the petroleum supply sector are $50,817 million, $64,384 million, and $66,785 million in the BAU, MPR, and CPD cases, respectively. As another example, input purchases by the industrial chemicals sector from the petroleum refining sector are $27,170 million, $29,550 million, and $25,150 million in the respective cases. Input purchases of the important conversion industries are significantly affected by the specifications of different energy/environmental policies.

Input Requirements

The principal objective of this study is to demonstrate the relationship between process models and the coefficients of an input-output table. To this end, matrices of direct and of direct plus indirect requirements are derived from the three transactions tables presented above. By definition, the direct requirements table is obtained by dividing each of the sector-specific inputs for a given sector by total input for that sector. (The resulting coefficients sum to near unity, given round-off error; no coefficient may be greater than unity and at least one column must be less than unity for the input-output structure to be stable (19, p. 23)). The direct plus indirect requirements table is derived from the direct requirements table by: (1) deleting the payments sector row, (2) subtracting the resulting square matrix from an identity matrix of the same dimension, and (3) inverting the matrix obtained from this subtraction.

The calculated direct plus indirect requirements matrices for each of the three cases are presented in Tables 10, 11, and 12. Each of these input-output structures corresponds to a specific structure of and solution to the equilibrium model of the energy microcosm. As can be seen, differences in the structures of the underlying equilibrium systems produce corresponding differences in the related input-output structures. (The direct requirements for the coal and petroleum supply sectors are the same in all cases because of the use of 1967 proportions.) Importantly, the matrices of direct plus indirect requirements are positive, thus satisfying the fundamental Hawkins-Simon condition (19, p. 27, and 5).

Table 7

Transactions Table, Business-as-Usual, 1967 dollars

	Coal Supply	Petroleum Supply	Petroleum Refining	Electric Utilities	Modeled Chemicals	Final Demand	Total Outputs
Coal Supply	1776.05	2.03	90.14	7888.50	2083.56	2210.72	14051.0
Petroleum Supply	4.22	2529.10	50817.00	3894.0	3498.43	41780.6	101,695.92
Petroleum Refining	155.97	223.28	428.94	6649.40	27836.04	23853.22	59146.85
Electric Utilities	292.26	787.55	0.0	936.21	2671.0	37388.57	42903.02
Modeled Chemicals	39.34	926.26	4122.36	537.22	9656.79	63772.86	79054.83
Payments Sector	11783.16	97227.70	3688.41	22997.69	33309.01		
Total Input	14051.0	101,695.92	59146.85	42903.02	79054.83		

Table 8

Transactions Table, Modified Price Regulation, 1967 dollars

	Coal Supply	Petroleum Supply	Petroleum Refining	Electric Utilities	Modeled Chemicals	Final Demand	Total Outputs
Coal Supply	1725.11	1.94	78.74	6811.20	1650.10	3380.01	13648.0
Petroleum Supply	4.09	2411.82	64384.0	4296.0	1500.72	24383.29	96979.92
Petroleum Refining	151.49	212.92	371.50	6875.20	30257.88	32878.27	70747.26
Electric Utilities	283.88	751.03	0.0	931.80	4445.17	36485.91	42897.79
Modeled Chemicals	38.21	883.0	2727.69	685.91	2569.77	68602.82	75507.4
Payments Sector	11445.22	92718.91	3185.61	23297.68	35083.76		
Total Inputs	13648.0	96979.92	70747.26	42897.79	75507.4		

Table 9

Transactions Table, Complete Price Deregulation, 1967 dollars

	Coal Supply	Petroleum Supply	Petroleum Refining	Electric Utilities	Modeled Chemicals	Final Demand	Total Outputs
Coal Supply	1760.12	2.02	36.0	7831.40	2058.86	2236.6	13925.0
Petroleum Supply	4.18	2519.21	66785.0	5816.00	4180.70	66785.0	101,298.3
Petroleum Refining	154.57	222.40	425.60	6754.50	25807.11	38651.52	75015.7
Electric Utilities	289.64	784.47	0.0	922.28	3500.03	39253.42	44749.84
Modeled Chemicals	39.99	922.63	4111.34	547.69	9508.32	66370.14	81500.11
Payments Sector	11677.51	96847.57	3657.76	22877.97	36445.09		
Total Inputs	13925.0	101,298.3	75015.7	44749.84	81500.11		

Table 10

Input-Output Tables, Business-as-Usual

	Coal Supply	Petroleum Supply	Petroleum Refining	Electric Utilities	Modeled Chemicals
		Direct Requirements			
Coal Supply	0.12640	0.00002	0.00152	0.18387	0.02636
Petroleum Supply	0.00030	0.02487	0.85917	0.09076	0.03379
Petroleum Refining	0.01110	0.00220	0.00725	0.15499	0.35211
Electric Utilities	0.02080	0.00774	0.0	0.02182	0.04425
Modeled Chemicals	0.00280	0.00911	0.06970	0.01252	0.12215
Payments Sector	0.83860	0.95606	0.06236	0.53604	0.42134
		Direct and Indirect Requirements			
Coal Supply	1.1501	0.0022	0.0071	0.2181	0.0485
Petroleum Supply	0.0194	1.0335	0.9243	0.2514	0.4238
Petroleum Refining	0.0188	0.0077	1.0441	0.1752	0.4285
Electric Utilities	0.0249	0.0087	0.0117	1.0304	0.0577
Modeled Chemicals	0.0057	0.0115	0.0927	0.0319	1.1785

Table 11

Input-Output Tables, Modified Price Regulation

	Coal Supply	Petroleum Supply	Petroleum Refining	Electric Utilities	Modeled Chemicals
			Direct Requirements		
Coal Supply	0.12640	0.00002	0.00111	0.15878	0.02185
Petroleum Supply	0.00030	0.02487	0.91005	0.10015	0.01988
Petroleum Refining	0.01110	0.00220	0.00525	0.16027	0.40073
Electric Utilities	0.02080	0.00774	0.0	0.02172	0.05887
Modeled Chemicals	0.00280	0.00910	0.03856	0.01598	0.03403
Payments Sector	0.83860	0.95606	0.04503	0.54310	0.46464
			Direct and Indirect Requirements		
Coal Supply	1.1494	0.0019	0.0045	0.1881	0.0394
Petroleum Supply	0.0205	1.0339	0.9629	0.2741	0.4379
Petroleum Refining	0.0188	0.0078	1.0295	0.1797	0.4386
Electric Utilities	0.0249	0.0088	0.0107	1.03	0.068
Modeled Chemicals	0.0047	0.0102	0.0504	0.0273	1.0581

Table 12

Input-Output Tables, Complete Price Deregulation

	Coal Supply	Petroleum Supply	Petroleum Refining	Electric Utilities	Modeled Chemicals
		Direct Requirements			
Coal Supply	0.12640	0.00002	0.00048	0.17500	0.02526
Petroleum Supply	0.0003	0.02487	0.89028	0.12997	0.05130
Petroleum Refining	0.01110	0.00220	0.00567	0.15094	0.31665
Electric Utilities	0.02080	0.00774	0.0	0.02061	0.04295
Modeled Chemicals	0.00280	0.00911	0.05481	0.01224	0.11667
Payments Sector	0.83860	0.95606	0.04876	0.51124	0.44717
		Direct and Indirect Requirements			
Coal Supply	1.1498	0.0021	0.0049	0.2070	0.0448
Petroleum Supply	0.0207	1.0338	0.9485	0.2923	0.4149
Petroleum Refining	0.0183	0.0072	1.0331	0.1682	0.3795
Electric Utilities	0.0248	0.0087	0.0108	1.029	0.0551
Modeled Chemicals	0.0053	0.0112	0.0741	0.0284	1.1608

For the most part, the input-output results demonstrate an increase in the share of the payments sector in response to greater stringency of noneconomic structural constraints. This shift logically results from higher imputed costs for these constraints in the equilibrium solution. Isolation of this effect was not an objective in the design of the scenario analyses available for this illustration but seems a meaningful and important subject for future study. (Computation of additional equilibrium solutions for this specific isolation was beyond the scope of work and resources for this study.) Ideally, two complete price deregulation cases--with and without noneconomic constraints--are needed to show definitively the increased payments sector and cost effects of noneconomic constraints. Such a definitive demonstration should provide a much sharper focus for analyzing the causes of the nation's rampant inflation and underinvestment in productive capital.

The calculated indications of structural change seem to underscore Miernyk's statement (12, p. 33):

> In making long-term projections, for a ten-year period, for example, one cannot assume that input coefficients will remain constant.

Structural changes induced by policy in the transition from petroleum to alternate energy sources are one set of significant factors affecting these coefficients.

For the central purpose of this study, the important observation is that the derived input-output structures are a viable and consistent means of summarizing the detailed information on technical change available in the process models and equilibrium solutions. Some of this detailed microeconomic structural information is inevitably lost in the transfer to an input-output structure, but the analytical significance of this aggregation loss depends upon the purpose of the application.

Balancing the Input-Output Table

This paper has presented a methodology for determining the technical input-output coefficients of production as direct functions of prices and as indirect functions of shadow prices induced by noneconomic constraints. An application of this methodology to the energy sector of the U.S. economy shows how these technical coefficients could be estimated for three different policy specifications, each with a significantly different economic structure. This demonstration naturally leads one to ask, "How could a consistent general price/quantity equilibrium be reestablished for the economy, if the structural change is limited to the energy sector?" Alternatively, "How will changes in energy policies affect market conditions in the energy sector and general economic conditions throughout the economy?" This important economic question has not been well answered to date. We do not provide a complete answer here, but describe how the methods of this paper may be directed to answering this question in future work.

To return to our two sector examples, the interindustry transactions values for each industrial sector in the energy microcosm may be summed to give the value of energy output used by the energy sector. The transactions values in the modeled chemicals row and the purchases of the energy sector from non-modeled industries in the economy (in the payments sector) may be summed to give the value of input purchases by the energy sector from the rest of the U.S. economy. These two sums provide the basis for determining the technical input-output coefficients of the energy sector column. Next, a summation is made across all energy subsectors of the purchases by the modeled chemicals sector and the non-modeled nonenergy industries (included in the microcosm final demand column); this summation gives the interindustry sales of the energy sector to the nonenergy sector. Similarly, sales of the modeled chemicals industry to itself and to the non-modeled nonenergy industrial sectors (in the demand column of the energy microcosm table) are added to the value of purchases by the modeled chemicals sector from non-modeled nonenergy industries (in the payments sector of the microcosm economy). These two additional sums provide the basis for determining the technical input-output coefficients of the nonenergy sector column. Interindustry transactions between non-modeled nonenergy industries must be known from other sources.

The final demand function for nonenergy commodities and the primary input supply functions for labor and capital are assumed to be known and available.

The following matrix notation is employed:

$c = \begin{pmatrix} c_1 \\ c_2 \end{pmatrix}$ represents the final demands for energy and nonenergy;

$A = \begin{pmatrix} a_{11} & a_{12} \\ a_{21} & a_{22} \end{pmatrix}$ represents the technical interindustry input-output coefficients;

$B = \begin{pmatrix} b_{11} & b_{12} \\ b_{21} & b_{22} \end{pmatrix}$ represents the technical coefficients of primary input use;

$r = \begin{pmatrix} r_1 \\ r_2 \end{pmatrix}$ represents the resource supplies of labor and capital;

$p = (p_1, p_2)$ represents the prices of the energy and nonenergy commodities;

$w = (w_1, w_2)$ represents the prices of primary inputs, labor, and capital; and

$x = \begin{pmatrix} x_1 \\ x_2 \end{pmatrix}$ represents the outputs of energy and nonenergy.

Then, following Intriligator (8) the primal L. P. problem for the economy is as follows:

<u>Primal</u>

$$\max_{x} \; p\,c$$

$$\text{subject to: } x = Ax + c, \quad Bx \leq r, \quad x \geq 0$$

Solution of this problem gives the allocation of outputs between energy and nonenergy that maximizes net national product, subject to limited resource availabilities.

The dual is as follows:

<u>Dual</u>

$$\min_{w} \; w\,r$$

subject to:

$$w\,B \geq p\,(I-A),$$

$$w \geq 0$$

Solution of the dual problem gives the prices of the primary factors that minimize resource cost (national income), subject to the competitive no-profit condition.

Because c and r are functions of p and w, an iterative procedure is generally necessary to find the commodity prices, p, and primary input prices, w, which satisfy the definition of a competitive equilibrium for the economy (given that the existence conditions are fulfilled). Using this iterative procedure to determine the fixed equilibrium point in prices and quantities completes a first set of iterations in evaluating how policy changes affect the structure of the energy sector, energy prices and quantities, and economy-wide prices and quantities. Another set of iterations is then required to evaluate how the technical coefficients of production for the energy sector are affected by these revised equilibrium prices for the economy. Completion of these evaluations identifies how the feedback loop between the economy to the energy sector affects market conditions in the energy sector. As needed, additional sets of iterations may be completed until the definition of a competitive equilibrium is fulfilled for the economy and the technical coefficient matrix, A, is consistent with the determined prices. This solution gives a consistently balanced input-output table for the economy. This balanced table is necessary to analyze salient productivity, inflation, investment, employment, resource, and environmental questions.

As studied here, the overall problem seems well suited to application of the Dantzig/Wolfe decomposition principle; see Dantzig (18, Chapter 23). The decomposition principle would partition the technology matrix for the economy into a common set of row constraints faced by the energy and nonenergy sectors. It would further partition the remaining constraints into (1) those for the energy sector that do not interact with the nonenergy sector and (2) those for the nonenergy sector that do not interact with the energy sector. This may be illustrated as follows:

$$\begin{pmatrix} A_{11} & A_{12} \\ A_{21} & A_{22} \end{pmatrix} \text{Common constraints of both sectors}$$

$$\begin{pmatrix} B_{11} & 0 \\ 0 & B_{22} \end{pmatrix} \begin{matrix} \text{Energy sector constraints only} \\ \text{Nonenergy sector constraints only} \end{matrix}$$

The detailed model for the energy sector would be used to find an equilibrium solution to the energy sector. This would give the technical coefficients of matrices A_{11}, A_{21}, and B_{11}. Use of this information and the matrices of technical coefficients for the nonenergy sector (A_{12}, A_{22}, and B_{22}) gives the following linear programming/input-output problem for the economy:

$$\text{Max } p_1 c_1 + P_2 c_2$$

s.t.

$$\begin{pmatrix} A_{11} & A_{12} \\ A_{21} & A_{22} \end{pmatrix} \begin{pmatrix} x_1 \\ x_2 \end{pmatrix} + \begin{pmatrix} c_1 \\ c_2 \end{pmatrix} = \begin{pmatrix} x_1 \\ x_2 \end{pmatrix}$$

$$B_{11} \; x_1 \leq r_1$$

$$B_{22} \; x_2 \leq r_2$$

$$x_1 \geq 0, x_2 \geq 0.$$

In solution of this L.P./I.O. problem, decomposition algorithms would allow:

- the variable coefficient part of the problem to be modeled as a "Generalized Linear Program" (see Dantzig (18, Ch. 22)),
- dynamic considerations to be incorporated in a multi-stage framework, and
- computational efficiency to be improved in finding an overall solution.

All of these gains need to be explored in further research.

References

1. Almon, C. Jr., M. B. Buckler, L. M. Horowitz, and T. C. Reimbold, 1985: Interindustry Forecasts of the American Economy, Lexington, Mass., Lexington Books, 1974.

2. Duffin, R. J., E. L. Peterson, and C. Zener, Geometric Programming--Theory and Application, New York, John Wiley and Sons, Inc., 1967.

3. Federal Energy Administration, Project Independence Evaluation System (PIES) Documentation, Vol. XIII, "Coal and Electric Utility Conventions for PIES." FEA/N-76/423, NTIS PB-265 824, September, 1976.

4. Griffin, J., "Long-Run Production Modeling with Pseudo Data: Electric Power Generation," Bell Journal of Economics, Spring, 1977, Vol. 8, No. 1, pp. 112-127.

5. Hawkins, D. and H. A. Simon, "Some Conditions for Macroeconomic Stability," Econometrica, 17 (July-October 1949, pp. 245-248).

6. Hoffman, K. C. and D. W. Jorgenson, "Economic and Technological Models for Evaluation of Energy Policy," Bell Journal of Economics, Autumn 1977, pp. 444-466.

7. Hudson, E. A. and D. W. Jorgenson, "Energy Policy and U. S. Economic Growth," Proceedings of American Economic Review, Vol. 68, No. 2, May, 1978, pp. 118-130.

8. Intiligator, M. D., Mathematical Optimization and Economic Theory, Prentice-Hall, Inc., Englewood Cliffs, N.J., 1971, Ch. 9.

9. Kim, Y. Y. and R. G. Thompson, Economic Supplies of Crude Oil and Natural Gas in the Lower 48 States, Houston, Gulf Publishing Company, July 1978.

10. Klein, L. R., "The Supply Side," Proceedings of the American Economic Review, Vol. 68, No. 1, March 1978, pp. 1-7.

11. Leontief, W., "Environmental Repercussions and the Economic Structure: An Input-Output Approach," The Review of Economics and Statistics, Vol. 52, August 1970, pp. 262-271.

12. Marshak, J., "Statistical Inference in Economics: An Introduction," Statistical Inference in Dynamic Economic Models, T. C. Koopmans (ed.), New York, John Wiley & Sons, Inc., 1950, pp. 1-52.

13. Thompson, R. G., J. A. Calloway, and L. A. Nawalanic (eds.), The Cost of Clean Water in Ammonia, Chlor- Alkali, and Ethylene Production, Houston, Gulf Publishing Company, 1976.

14. Thompson, R. G. and J. A. Calloway, and L. A. Nawalanic (eds.), The Cost of Electricity, Houston, Gulf Publishing Company, 1977.

15. Thompson, R. G., J. A. Calloway, and L. A. Nawalanic (eds.), The Cost of Energy and A Clean Environment, Houston, Gulf Publishing Company, 1978.

16. Thompson, R. G., F. D. Singleton, Jr., J. C. Stone, and Y. Y. Kim, "Energy Supplies, Demands, and Prices for the U.S. Economy in 1985," National Energy Act, Hearings before the Subcommittee on Energy and Power of the Committee on Interstate and Foreign Commerce, 95th Congress, 1st Session, Serial No. 95-26, June 1, 1977, pp. 399-456.

17. U.S. Department of Commerce, Input-Output Structure of the U.S. Economy: 1967 Volume 1, Transactions Data for Detailed Industries, 367 sectors, A Supplement to the Survey of Current Business, 1974, Supt. of Documents, Government Printing Office, Washington, D. C. 20402.

18. Dantzig, G. B., Linear Programming and Extensions, Princeton, N.J., Princeton University Press, 1963.

19. Miernyk, W. H., The Elements of Input-Output Analysis, New York, Random House, 1965.

Part 3

Transfer and Alternatives to Transfer for Large-Scale Models

Introduction: Part 3

Large-scale energy models are being developed and used in the debate over national energy policy. Because these models are being used and because they are large and complex, modelers and users need improved model credibility. A key factor in improving model credibility is the enhanced ability of peers to critique the models. Model evaluation requires good-quality documentation and other characteristics so that analysts can reproduce the work of the model builder and can compare the work with alternatives. Further, the models must be simple and clear so that users can understand why certain results are forthcoming from the models and will know the important simplifying assumptions upon which models are built and used.

The first four chapters of Part III relate to the issues of transfer, alternatives to transfer, and model comparison as methods to achieve fuller understanding of model behavior by peers and users. The fifth chapter in Part III is an overview of current energy modeling as compared to an ideal modeling approach. Milton Holloway argues in chapter nine that there currently is no good substitute for the transferability of models in order to achieve a high measure of credibility. He summarizes the Texas/DOE experience of the Texas National Energy Modeling Project (TNEMP) in making a case for transferability, and concludes that proper institutional arrangements for developing and using models will internalize incentives for providing good model attributes; the degree of attention given to documentation, testing, and peer reviews will be appropriate to the value of the information.

Transferring a large model not necessarily designed with transfer in mind entails both cost and effort. Roger Glassey describes in chapter ten the effort required by DOE to accommodate the TNEMP. He also points out the limitations of transfer to achieve an adequate level of understanding on the part of competent but untutored recipients of a large model. A model cannot be entirely separated from the modeler.

There are other approaches to achieving most of the benefits of large model transfer for establishing credibility. Bill Hogan describes in chapter eleven the activities of counter-modeling, use of mini-models, remote access, and model comparison as practical alternatives to actual transfer. He argues that these alternatives are less expensive and, when coupled with a better view of the role of models, provide acceptable alternatives to transfer.

Institutional arrangements can be structured to bring together modelers and users of models to improve the understanding, credibility, and use of models. Such an institutional arrangement is the Energy Modeling Forum described in chapter twelve by Jim Sweeney. Requiring modelers to operate their models with like data and assumptions helps identify errors, clarify disagreements, and provide guidance for model selection by users.

Models tend to be oversold in their capability to forecast and provide sound analytical results. Charles Holt reviews in chapter thirteen the current state-of-the-art in large-scale energy modeling as compared to an ideal modeling approach. He also suggests the creation of institutions, which he calls research centers for modeling, to aid in the development and use of models.

Although the papers in Part III are certainly not exhaustive, they focus attention on a major area of concern in the development and use of large-scale energy models. All the authors recognize the need for activities that increase the credibility of models and offer important perspectives to an emerging debate and early attempts to do something about the problem.

Milton L. Holloway

9. The Importance of Transfer in Understanding Large-Scale Models

Introduction: The Current Status of Large-Scale Modeling

The primary purpose of this paper is to review the costs and benefits of a particular large-scale model transfer and evaluation experience so that comparisons may be made with alternatives to transfer. Before discussing the specifics of the model transfer and evaluation, however, it is useful to place the experience in the context of current large-scale models.

Large-Scale Models Used in Policy Work

Large-scale models are a product of our times and are growing more important in formulating government policy. Policymakers are using large-scale models, having no apparent alternative, but at the same time they are skeptical of the reliability of forecasts and calculations made by models. In addition, people generally accept results of computer analyses, which seem to them to represent the epitome of technological solutions to problems. Analyses are seen as more believable if they are based on computer technology.

Examples abound of the recent use of large-scale econometric models in the debate over national economic policy. The most prominent among these large-scale models used in economic policy planning by government include the DRI, Chase, and Wharton models.

In addition to government's use of econometric models in economic planning, new computer modeling is used for energy policy planning and analysis. Some of the better known of these modeling efforts include the Department of Energy's (DOE) Project Independence Evaluation System (PIES), now known as Midterm Energy Forecasting System (MEFS), the Brookhaven Energy Reference System, the Stanford University PILOT modeling effort, and the Baughman-Joskow Regional Electricity Model.

In energy policy analysis these models have been used most recently for projecting the world crude oil situation and the impact of U.S. Government conservation and production incentives on the U.S. economy, in producing annual energy outlook reports for the U.S., and in the national debate over natural gas pricing during the discussion of President Carter's National Energy Plan (NEP). Secretary Schlesinger, in discussing the prospects for improvement in the energy outlook as a result of the National Energy Act, gave quantitative answers to increases in production and decreases in foreign imports but promptly added that the models producing these estimates "aren't worth much" (1). Other examples exist, illustrating that models are being used but with a great deal of skepticism; users are not very confident of their reliability.

The major task before us is to improve the usefulness of models and the judgments of professionals involved in public policy analysis. At issue are the procedures by which the reliability of large-scale models can be established and made transparent--to distinguish between the influence of professional, subjective judgments and the influence of objective information reproducible by others.

Congress became concerned over the reliability of data, model results, and procedures of documentation used by the Federal Energy Administration and created a Professional Audit Review Team to audit FEA's activities and to report to Congress annually. Such audits may help, but I do not believe that such an approach solves the basic problem.

Testing, Validation, and Verification Activities Inadequacies

A number of institutional and professional forces affect current large-scale model development to the detriment of needed validation and verification. Communication between modelers and users is also typically absent, resulting in lack of clarity in model attributes and, more importantly, in lack of relevancy of model output to the problems of the user (2).

Three factors account for poor performance in model testing and validation. First, sponsoring entities fail to provide time, good management practices, and adequate funding for model verification and validation. Second, professional interests and rewards are much stronger for model building and use than for model testing; there is little reward for model evaluation. Third, there is a strong demand for use of untested, favorable model results by the political system. These conditions are responsible for a number of formal and ad hoc attempts to correct the situation, and some believe that the problems are even spawning a new discipline of model evaluators.

The Comptroller General of the United States recently issued a report concerning ways to improve management and development of federally funded computerized models (3). The General Accounting Office (GAO) identified 519 federally funded models developed or used in the Pacific Northwest. Fifty-seven of these models, costing $21 million to develop, were studied. The study found problems with federally funded modeling in three areas: (1) inadequate management planning, (2) inadequate management commitment, and (3) inadequate management coordination. In the area of management planning, the study identified a lack of clearly defined problems to be modeled, inadequate data to make models functional, insufficient funds to complete the models, a lack of provision for updating models for future use, a lack of provision for evaluating the models, and a lack of requirement by management for documentation.

In the area of inadequate management commitment, the GAO study found problems due to a lack of active participation in planning of model development by management and a lack of management understanding of modeling techniques and applications. GAO found two major problems in the area of inadequate management coordination. Management did not monitor model development on a continuous basis and sponsoring managers did not coordinate the model development with the developer. As a follow-up to this study GAO is preparing a booklet on guidelines for model evaluation (4). The booklet is intended to be a guide to decision makers for assessing the reliability of models.

Professional incentives encourage modelers to develop and use models but there are few rewards for validation and verification. Publication in professional journals is a good indicator of the imbalance of professional incentives (Table 1). A count of published articles in the American Economic Review, Operations Research, and the Bell Journal of Economics over the last six-year period illustrates the point. The results are as follows:

Table 1. Survey of Professional Journals in Economics

Journal	Total Number of Articles	Articles On Model Development or Use	Articles On Verification or Validation
AER	464	221 (47.6%)	0
OR	345	159 (46.1%)	3 (0.9%)
BJE	228	121 (53.0%)	0

These tabulations are only indicators but it seems clear that verification and validation are not topics that receive space in professional journals. It is also clear that models are popular topics (5).

Models developed for use in the political policy-making process are not completed and used in isolation. Models and their results become the tools of experts near the center of political power. Experts deal in a scarce commodity, knowledge, which includes not only the knowledge to which they have access, i.e., their expertise, but, more importantly, the information they obtain and generate (6). Because of their control of scarce resources experts derive considerable power; this power is enhanced the nearer the expert is to the seat of political power, which is vested principally in elected officials in the United States (7). But access to the center of power by experts is always limited since many issues command the attention of political leaders. Therefore, timeliness of information from data bases and models will always be important; hence, the strong tendency to use preliminary information or information from untested, unvalidated, and unverified models. Information is highly valued in the policy process and timeliness is often the key determinant of its value. Strong countervailing incentives will be required to offset these demands and to increase model testing.

Political and Professional Responses to Modeling Weaknesses

The absence of adequate testing, validation, and verification in modeling development and use has created both political and professional interest in correcting the situation. The GAO studies cited previously are examples of political recognition of a problem and suggested regulatory approach to a solution. The Electric Power Research Institute now sponsors two ongoing forums for model comparison. One is the Energy Modeling Forum (EMF) at Stanford University, and the other is an independent model assessment activity, the Model Assessment Laboratory at MIT. These and other forms of model analysis are summarized in a recent article by Greenberger and Richels (8).

Greenberger and Richels describe two kinds of model analysis that typically begin to appear during the later stages of model development, that is, during a model's use in policy debate. One type of analysis is that which focuses primarily on the model; the other type focuses more on policy issues and uses of the model (9). The authors define a taxonomy for model analysis by setting up a two-way classification system contrasting the main party in an analysis against the motivation and setting for the analysis (10). The Greenberger/Richels chart of taxonomy for model analysis is presented in Figure 1.

Figure 1. Two-way Classification of Model Analysis

Main Party	Motivation and Setting		
	Natural	Ad Hoc	Institutionalized
First Party Model Developers			
Second Party Model Users			
Third Party Model Analysis			
Joint Effort Mixed Group			

Source: Greenberger, Martin and Richard Richels, "Assessing Energy Policy Models: Current State and Future Directions," Ann. Rev. Energy, 4:467-500, 1979, p. 475.

In the first stage of model development analysis is done by the modelers themselves (this type of activity is done as a natural part of the modeling development), by ad hoc groups and workshops, and by institutional self-assessment and standards. The second group of model analysts are users, and work will be done as a normal part of staff activities, by consultants reporting to the user on an ad hoc basis, or by ongoing user review that has been institutionalized. The third category of analysts is a third-party situation. These groups may function as a natural, spontaneous peer review group, an ad hoc, organized review group, or as institutionalized model assessment laboratories. The fourth category of analysts is a mixed group, containing some or all of the first three types and functioning as a natural "marketplace of ideas," ad hoc discussion groups and studies, or as an institutionalized ongoing forum (11).

Greenberger and Richels categorized the Texas National Energy Modeling Project as a mixed group, ad hoc model analysis exercise. Greenberger and Richels' taxonomy is a useful reference system for discussing the Texas National Energy Modeling Project, and it will be referred to again.

Critiques by the four types of analysts in one of three settings use a number of differing approaches, but such evaluations may be done either with or without hands-on operation of the model being evaluated. The central purpose of this paper is to point out the benefits and costs of a large-scale model critique with hands-on operation completed in an ad hoc setting by interested users and modelers. This critique was sponsored by an energy policy office in Texas that reflects the attitudes of Texans affected by national energy policy.

The Texas National Energy Modeling Project: An Experience in Large-Scale Model Transfer and Evaluation

During the winter and spring of 1978 the Texas National Energy Modeling Project (TNEMP) was conceived and organized. During the fall of 1978 and continuing through the summer of 1979 the Midterm Energy Forecasting System (MEFS), formerly the Project Independence Evaluation System (PIES), was transferred to and operated on a Texas computer, an evaluation of all principal model components of the system was completed, and a workshop was held in Washington, D.C., to discuss the findings of the evaluation with representatives of the Energy Information Administration (12).

Project Purpose

The first purpose of TNEMP is to provide an independent evaluation of the Energy Information Administration's (EIA) MEFS. The evaluation will guide users of MEFS concerning the level of confidence one may have in the results of the models used for government energy policy analysis. The evaluation is critical, in the best sense of the term, and it suggests improvements in the model structure and in procedures used by the EIA for increasing model credibility.

The second purpose of TNEMP is to provide recommendations to the Texas Energy Advisory Council (TEAC) (13) for the Council to maintain a national modeling system to evaluate Texas impacts consistent with national modeling. As a result, we have first-hand experience with MEFS, as well as with Data Resources, Inc.'s (DRI) macro and energy models. We can compare these models, their structure and results with various Texas models at TENRAC, the University of Houston, and The University of Texas to assess their relative usefulness for energy policy analyses and their possible joint use or integration. TNEMP participants have experience in still other modeling efforts, as well as in model evaluation, institutional arrangements for housing models, and procedures for maximizing resources for successful model development, use, and credibility. TNEMP is thus well suited for the second purpose.

Evaluation of national energy policy decisions impacts on Texas must necessarily be done within a consistent national framework, since a major portion of the nation's energy production and processing, as well as corporate energy management for the nation as a whole, is in Texas. A significant fraction of the nation's energy, especially natural gas, is also consumed in Texas. Texas-based energy analyses must, therefore, be national to reflect Texas' interests and its role in the nation's energy future accurately. To be useful in the national energy policy process, Texas-based analyses must be based on a credible national modeling and analysis framework.

A third purpose of TNEMP is to focus attention on the current uses, practices (potential as well as current abuses), and potential dangers of modeling for developing public policy and clarifying important issues to the citizens of a democratic society. The increased reliance of policymakers, high-level advisors, and the voting public on expert opinion, based in part on large-scale data bases and models, in an era of complex problems and policy prescriptions, requires that policymakers and high-level experts gain a better understanding of current modeling reliability and practices. TNEMP is intended to help achieve this goal.

Two additional factors are important side effects of TNEMP. First, we have advanced the art of model evaluation by achieving, among other things, the transfer and operation of a large and complex model. We hope that others will benefit from both our successes and our mistakes, for model evaluation is far from being a well-developed discipline. Second, we hope that some insight has been gained into the appropriate institutional setting for model evaluation. Current institutional arrangements for building, operating, and maintaining large models lack procedures aimed at validation of relations, assumptions, and data, verification of model outputs, and model documentation. We seek model evaluation institutions and/or incentives to bring about a commitment on the part of model developers to validation and verification, clarity, and workability as model attributes. Third-party evaluation can increase the probability of a model being accepted and it can enhance the transfer of scientific, technological, and statistical data to the policy decision-making process.

Organizational Structure and Procedures

The organizational structure of the study consisted of four primary groups. First, I provided project direction, and the staff provided coordination, materials, and other support. Second, the National Advisory Board provided advice on procedures, suggestions on methodology, evaluation of the Analysis Team results, and general recommendations for a Texas national modeling capability. Third, the Analysis Team evaluated MEFS, suggested improvements and alternatives, and made specific recommendations for Texas' maintenance of a national modeling capability reflecting the impact on or by Texas. Fourth, the Supporting Institutions provided support by endorsing the objectives of the study, making review and

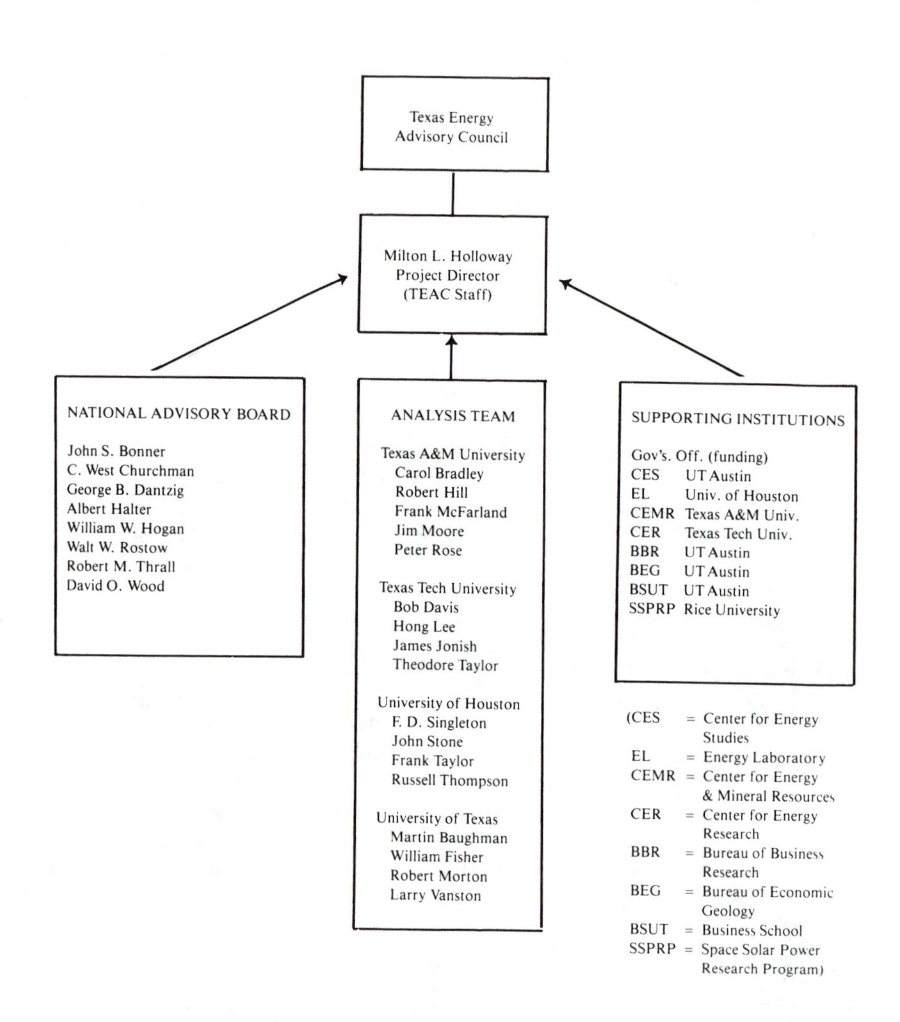

Figure 2. Organizational structure of the Texas National Energy Modeling Project.

comment, making data and facilities available, and funding the study. Figure 2 illustrates the working relationships of the four groups.

Meetings were held periodically to report the progress of various studies to the National Advisory Board, other members of the Analysis Team, and Supporting Institutions. These meetings provided the opportunity for refining project objectives, defining evaluation criteria, sharing reference material, identifying weaknesses in the project study design, interacting with DOE personnel, and providing overall guidance to the project director and principal investigators.

In order to document the project objectives and procedures, the National Advisory Board was asked to write an evaluation of the study to be published with the final report. To encourage maximum intergovernmental cooperation, EIA was invited to comment on the project with the assurance that the remarks would also be published with the final report; EIA comments are included in the final report. So that key decision makers would be familiar with the important findings of the project, a meeting was arranged between members of the National Advisory Board, officials of EIA, and the Lieutenant Governor of Texas, who served as Chairman of TEAC. Results of a final workshop involving TNEMP and EIA were published with the final report to summarize the agreements and disagreements at the end of the project.

Evaluation Criteria

The reliability of information from the EIA large-scale energy model known as MEFS, currently being used for forecasting, policy evaluation, and policy analysis, was our primary concern in this study. Determining reliability is a difficult task; much has been written recently on the topic but little is well-defined and clarified. We made interpretations concerning the process and measures to be used in assessing reliability. According to the usual definition of reliability in the scientific sense, modeling results should be unambiguous, reproducible, and transferable. Such characteristics need further definition, however, since we are dealing with a system involving the behavior of people (the economy with emphasis on the energy sector) and applying models to describe alternative worlds, some of which will never exist for testing purposes. We also need to be specific with respect to the user of the information: in our case, users will be both other modelers and professionals as well as policy decision makers and the public at large. It is also essential to distinguish between using the models for raising issues or for resolving issues. Models used for raising issues may rely more on hypotheses for their formulation, whereas models used and designed for resolving issues will necessarily have to be based on accepted theory and/or laws. The MEFS model is used primarily by policymakers and the public at large for resolving energy issues. For that reason the evaluation criteria we selected for TNEMP were specifically defined.

No consensus now exists on either the appropriate set of criteria for evaluating models or the definitions of commonly used terms in model assessment literature and discussions. Various discussions of model assessment criteria may be found in recent publications. Four common themes run through the discussions and were the basis of our evaluation criteria in TNEMP: workability, clarity, validity (coherence), and verifiability (correspondence) (14).

The four themes commonly discussed in connection with model assessments are interrelated. First, if the model is to serve any useful function other than the satisfaction of the modeler's intellectual curiosity, then it must identify issues, suggest a resolution of issues, or in general produce results that relate to the practical problems of the user. We have interpreted this theme as our criterion of workability. In the context of the energy model evaluated, we interpreted workability to mean that the model could provide information on such topics as the inflationary impacts of decontrol policy, the impacts of conservation and decontrol on the import levels of crude oil, the effect of the combination of energy policies on international trade and the value of the dollar, the impacts of deregulation, coal conversion, and other such policies on economic growth, and perhaps the regional distribution of economic growth as a result of national policy. So in the context of our model evaluation effort the model is workable if it raises important policy issues in these areas, points toward the resolution of important issues, or provides explicit information pertinent to an issue. To be workable a model must provide relevant information for the resolution of a particular practical problem.

Second, a model must be clear or unambiguous. Clarity must be defined in terms understandable to other modelers (who, for example, may serve as advisors or interpreters to laypeople and/or policymakers) and in terms understandable to laypeople and/or policymakers, who are the users. For our model evaluation, the model's behavior and its results must be translatable into various supply-and-demand representations of the economy, understandable to economists, and intuitively comprehensible by laypeople who want to know the inflation, income, tax, energy costs, employment, and economic growth implications of the policies being analyzed. This understanding is embedded in our use of the criterion of clarity.

Third, a model should be logically consistent or coherent. For the model to be coherent it must be consistent with the underlying logic or theory upon which it is built. Further, the model must do what the modeler says it will do according to its design. The model will not be accepted by the disciplines if the basic theories are violated or if new or modified theory is required of a model designed to resolve policy issues. To develop a model representing a market economy in the absence of accepted economic theory is to challenge the basis for the economics profession; and use of the model will, therefore, have a much different effect than a model based on well-accepted economic theory. Coherence at least partially depends upon understanding the disciplines best represented by the current body of theory. In terms of our model evaluation (since MEFS is explained by EIA to be a model of the U.S. energy market and macro economy) we should be able to verify that increased prices for oil and gas simultaneously bring on increased supplies, reduced demands for oil and gas, and increased use of alternative energy sources. The model must behave in a manner consistent with the economic theory of markets in the context of the U.S. domestic economy; if they do not, the theories or model require modification. This theme is consistent with our criterion of validity or coherence.

Fourth, there is the question of the relationship of the model to "reality." It is tempting to state the fourth criterion (verifiability) in such phrases as "the model ought to represent reality." Such a criterion works reasonably well for model airplanes or planetarium models of the solar system, because we can observe both the reality (real airplanes or real planets) and the model results or consequences and identify some correspondence. But the reality of an "energy system" and "the economy" cannot be observed in such a simple manner. Instead, there are data banks, usually but not always put together by the disciplines, that are at best major abstractions and aggregations. It is reasonable, however, to expect that the model results should agree reasonably well with the accepted data; but if it does not, then some satisfactory explanation should be forthcoming, such as, the data are incorrect or incomplete, or the models require modification. The model consequences or results must correspond to reality in this sense. This theme meets our criterion of verifiability or correspondence (15).

In TNEMP we have considered this set of four criteria from the modeler's, the model assessor's, and the final user's point of view. We suggest that the responsible modeler should also use such criteria. The criteria are also useful to the model assessor. MEFS is judged by our group on its workability in solving a user's practical problems; to a lesser extent the models were evaluated on the basis of their ability to raise or to help resolve important issues from an analyst's, interpreter's, or advisor's point of view. We have paid particular attention to the <u>apparent</u> intended uses of the model (as evidenced by its historical applications). Validity was judged in our project from the analyst's and final user's point of view. Extensive comparisons were made with current economic theory, since the model attempts to represent the market economy. We also paid attention to MEFS's relationship to geology and engineering, which must be coherent with accepted theory in those fields. Validity was judged by model assessors by comparing the model relationships and behavior with logical deductions based on the theoretical underpinnings of the modeling system. Verification work was not documented by the modelers. Work on verification by the model assessors was not extensive. It was impossible to operate the models comprehensively to obtain results for time periods when data are available, exclusive of that used to estimate model parameters.

Project Costs

The total monetary cost to Texas of the TNEMP model evaluation was roughly $300,000. Of this total, $225,000 was for professional services, materials, and travel for the Analysis Team and the National Advisory Board. Approximately $25,000 was spent for computer charges and software rental. The remaining $50,000 is accounted for by in-kind services of the Project Director and TEAC staff in the management of the project. In-kind costs (and possible small amounts of direct costs) were also incurred by EIA, but this information was not available to us.

The project involved approximately one year of effort by 26 professionals in TNEMP in varying degrees of participation. The study required roughly 6.5 man-years of effort at the Texas end of the project plus whatever effort was expended by EIA.

TNEMP commissioned 11 studies in four groups: (1) computer operations of the Midterm Energy Forecasting System (MEFS)--study 1; (2) supply modeling--studies 2, 3, 4, 5, and 6; (3) processing and transportation modeling--studies 7, 8, and 9; and (4) demand modeling with macroeconomic interface modeling--studies 10 and 11. These studies encompass physical relationships in engineering and geology, industrial and consumer economic behavior, and the economics of conditional market equilibrium, since this is the domain of MEFS.

Because MEFS is a large and complex modeling system, developed over several years, costing many man-years of effort and several million dollars, it is not practical to evaluate every aspect of the model and the data. Therefore, the initial study design focused on evaluation, using hands-on operation of the computer, of (1) crude oil, natural gas, and coal satellite models; (2) transportation, electric power, and refinery (and synthetics) sector representations in the integrating linear programming model; (3) the macro-/microeconomic interface in all its aspects and the specific formulation of the demand models with actual operation of the Data Resources, Inc. (DRI)/MEFS interface on the computer; and (4) the integrating model. The 1977 National Energy Plan (NEP) version of MEFS was transferred to Texas for purposes of the evaluation.

Figure 3 shows the parts of the modeling system evaluated in this study. The NEP version of MEFS was "brought up" on the Texas A&M University Amdahl computer, and the principal investigator in charge of the system provided computer and operating support for other members of the team. Other supply data and related calculations, including uranium for nuclear power production, solar and other new technology contributions, data and programs for estimating conservation impacts on the demand side, and data and calculations specifying non-crude oil import levels, were ignored in the evaluation.

A mid-course correction in the study set priorities for the hands-on operation of the computer to emphasize: first, testing the oil and gas supply model behavior with impacts on the integrating model solutions; second, operation of the coal supply model with impacts on the integrating model solutions; and third, operation of the system to test macro/micro model interface results and the impacts of errors or variation in demand model parameter estimates. During the October 1978 National Advisory Board meeting and discussions with Energy Information Administration (EIA) personnel, it was determined that the NEP version of MEFS could not be verified by either EIA or TNEMP, nor could TNEMP operate it intelligently, due to non-existent documentation. Therefore, it was agreed that TNEMP would focus instead on the 1977 Administrator's Annual Report version of the model. This version was accordingly transferred and operated on the Texas A&M computer for TNEMP evaluation.

Project Benefits

Major benefits of the TNEMP exercise have and will continue to accrue directly to both TEAC (now TENRAC) and EIA and indirectly to modelers and model users at large. The experience will also benefit those more directly engaged in building institutions and incentives to promote model testing, validation, and verification as integral parts of model development projects. Finally, further discussion on issues of model assessment criteria and a better understanding of the role of models in the policy process spawned by this and other studies are of general benefit to both analysts and policymakers.

More specifically, the major benefits of TNEMP jointly to the Texas Energy and Natural Resources Advisory Council and DOE/EIA are:

1. establishment of a dialogue on energy modeling relevant to current policy issues; this dialogue is an important source of learning for both groups;

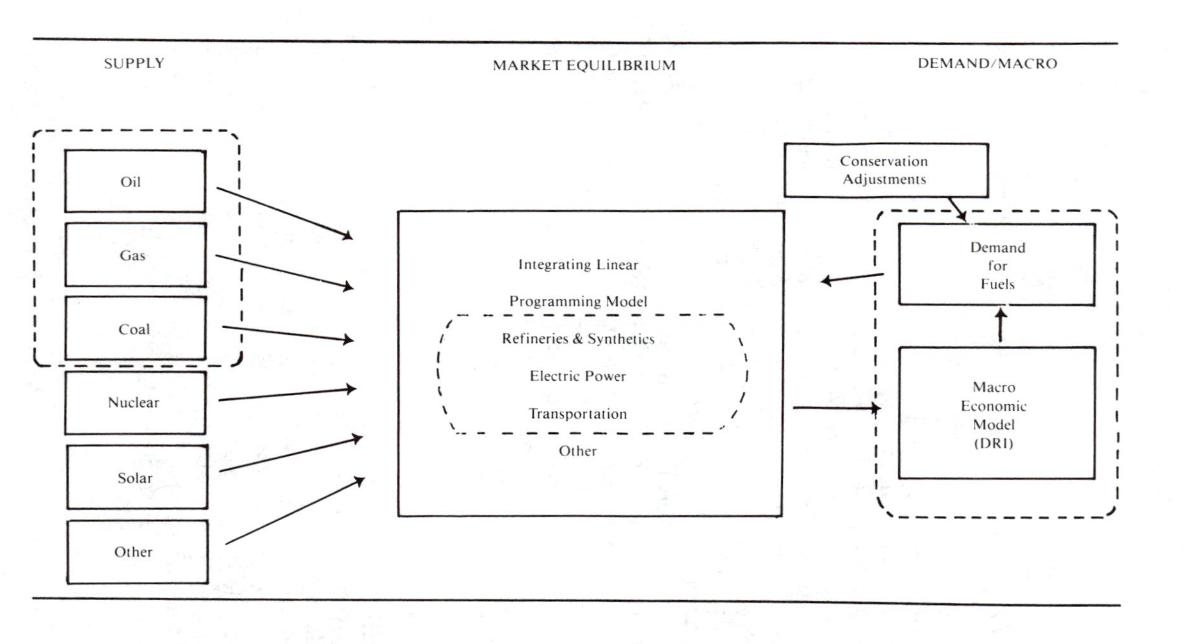

Figure 3. Simplified Overview of Midterm Energy Forecasting System Components. Circled components indicate parts of MEFS evaluated in this study. The oil, gas, and coal supply models were operated on the computer using selected parameter variations for sensitivity analyses; the demand model was mechanically operated on the computer but sensitivity analyses were done by ad hoc adjustments to demand model parameters; existing results of the DRI model were used to test the sensitivity of the macroeconomy impacts of energy market shocks; and the integrating model was operated many times to incorporate pair-wise comparisons of market equilibrium solutions from changes in the demand, oil, gas, and coal supply models.

2. the successful transfer to Texas and operation of MEFS; this adds credibility to both organizations' use of models in policy and analysis work;

3. the clarification of the staff time and monetary requirements for large-scale model transfer;

4. the institutionalization of an annual Texas review of EIA's annual report to Congress now included in TENRAC's statute; this will maintain the ties between the two groups.

The major benefits of TNEMP primarily to the Texas Energy and Natural Resources Advisory Council are:

1. an increased understanding of the major strengths and weaknesses of the DOE/EIA Midterm modeling capability; we will be better able to determine advisable and inadvisable uses of MEFS and to provide alternative analyses of major issues, especially concerning impacts of policy on the State of Texas and the surrounding region;

2. an institutional arrangement for an improved working relationship between TEAC and major Texas universities, created by the formulation of TNEMP;

3. a current and common base of understanding of the state-of-the-art in energy modeling on which to build an improved modeling system for use in Texas.

The major benefits primarily to DOE/EIA are:

1. increased justification for a new program emphasis on model documentation and verification already recognized as a major weakness as well as the opportunity in ongoing work with TEAC to test this system;

2. identification of specific documentation weaknesses and errors in various algorithms and lack of clarity in procedures for using MEFS;

3. use of the TNEMP evaluation studies in the design of new or modified models.

Future Work

TENRAC is currently planning a new program to refine its own modeling capability for assessing the regional impacts of national energy policy proposals used in the decision-making process. No other institutional entity can complete such analyses adequately. Such modeling will take full advantage of the results of TNEMP and the available current version of MEFS.

TENRAC will continue to transfer current "living versions" of MEFS to Texas for operation and maintenance as a service to other users, providing access to the modeling systems and data bases, and as an aide to our own modeling development and the requirement to evaluate the EIA's annual report. Texas has institutionalized an annual version of EIA's Administrator's Annual Report to Congress, reviewing alternative projections and outlook for the benefit of Texas and the nation. This mandate is included in the statutory requirements of TENRAC. Transfer of last year's version of MEFS is now underway.

Referring to the Greenberger/Richels taxonomy, TNEMP is a joint effort, institutionalized model analysis. It is a joint effort because it involved model developers (both in Texas and Washington), model users (TENRAC, representing the views and interest of Texans, as well as advisors on energy policy), and third-party interests (National Advisory Board). It is institutionalized because of TENRAC's responsibility to the Texas Legislature for an annual evaluation of EIA's annual report to Congress.

The Role of Transfer in Establishing Credibility of Large-Scale Models

There may be better and more efficient means of achieving the benefits of the type attributed to TNEMP in the previous section. Bill Hogan's paper deals with the topic of alternatives to transfer. Such alternatives are not in common practice, however, and it will be difficult to achieve some benefits through alternative approaches. Three such benefits are discussed here.

First, the actual transfer of models with hands-on operation demands a level of attention to detail by both modelers and model evaluators that is not easily reproducible. Interaction among the individuals involved provides a valuable source of information leading to better modeling decisions by the developer. Further, the evaluator, who is also an interested (affected) party, is the better source of information; disinterested parties by definition lack the interest to identify important practical problems in models and model applications.

Second, the transfer or the prospect of a requirement to transfer a model is in itself an incentive for modelers to document their work that is not provided by alternative processes. Large computer models create too many opportunities for too many participants to hide errors and weaknesses, and to pay inadequate attention to detail. The experience at the MIT Model Assessment

Laboratory documents the fact that the hands-on operation may often lead an evaluator to opposite conclusions concerning model behavior or the relative importance of some weakness (16).

Third, projects like TNEMP are sufficiently powerful to restrain experts from using information from models before validation and verification are complete. The current institutional arrangements for the use of models in Washington, D.C. seem to require this type of countervailing power. I hasten to add, however, that there is a better solution.

Models properly conceived and developed within the context of a decision-making unit contain an internal mechanism for ensuring proper attention to workability, clarity, verifiability, and validity as model attributes. The attention given to documentation and verification will be relative to the value of the information the models provide. A decision-making unit that is required to bear responsibility for its decisions and actions will discipline experts to take care in providing models that are relevant, clear, and documented.

The decision-making unit will be composed of an investigative unit where models, data bases, and other information tools are brought to bear on problems. Problems are identified jointly by investigators and decision makers. Interaction with those to be affected by decisions of the unit is also sought; that is, reactions from those affected by policy are obtained before decisions are made.

When models are developed with specifically defined problems in mind, agreed upon mutually by the decision maker and the modeler with feedback from those to be affected by decisions, outside regulation or policing is less needed. Models will be documented and clear if the developers know that they will be required to explain the models to a decision maker who wants to understand expert information from models. Modelers will design realistic models if they are required to deal with those who will be affected by the decisions influenced by the models.

Our attention should be directed more to institutional arrangements to internalize the modeling process within the context of decision-making units, rather than to creating independent, "objective" modeling agencies and third-party evaluators. The design of the second phase of TNEMP (to be known as the Texas Energy Policy Project) will have these institutional characteristics.

Summary and Conclusions

TNEMP was created to transfer and evaluate the Energy Information Administration's (EIA) Midterm Energy Forecasting

System (MEFS, formerly PIES) and to make recommendations on further modeling to the Texas Energy Advisory Council. The transfer was a rare experience in large-scale modeling transfer and was a successful venture. The cost to Texas of this study was approximately $300,000 and one year of effort in varying degrees of participation by 26 professionals in TNEMP, roughly 6.5 man-years of effort. Benefits from the transfer accrued to both EIA and the Texas Energy Advisory Council (now the Texas Energy and Natural Resources Advisory Council), and progress was made in the discipline of model evaluation.

The transfer of models seems to be a requirement for ensuring adequate attention to workability, clarity, verifiability, and validity as model attributes under current institutional arrangements in the public sector. A better long-term solution, however, is the development of models in the context of decision-making units that internalize the incentives for providing the above model attributes. The attention given to these model characteristics will be appropriate to the value of information provided by the models.

Footnotes

1. Bachman, W.A., "What the Industry Can Expect Now from the Department of Energy," The Oil and Gas Journal, November 13, 1978, p. 156.

2. Modeler/user interface weaknesses and poor performance in validation and verification work are closely related.

3. Comptroller General of the United States, Ways to Improve Management of Federally Funded Computerized Models: Report to the Congress, LCD-75-111, U.S. General Accounting Office, Washington, D.C., August 23, 1976, p. 44.

4. U.S. General Accounting Office, Guidelines for Model Evaluation, PAD-79-17, Exposure Draft, Washington, D.C., January 1979.

5. Closer checking, of course, may show that modeling articles contain reports of validation and verification tests associated with the model being reported, but the topics do not show up in article titles.

6. Benveniste, G., The Politics of Expertise, Boyd and Fraser Publishing Company, San Francisco, 1977, p. 145.

7. Ibid., p. 145.

8. Greenberger, M. and R. Richels, "Assessing Energy Policy Models: Current State and Future Directions," Ann. Rev. Energy, 4:467-500, 1979.

9. Ibid., p. 471.

10. Ibid., p. 475.

11. Ibid., p. 475-476.

12. Holloway, M.L., ed., Texas National Energy Modeling Project: An Experience in Large-Scale Model Transfer and Evaluation, Academic Press, Inc., New York, 1980.

13. The organization is now known as the Texas Energy and Natural Resources Advisory Council.

14. Selection of these criteria provides a definition of objectivity. The definition of objectivity will vary according to whether one adopts a positivist, normativist, or pragmativist philosophical position. See Holloway, M.L., ed., Texas National Energy Modeling Project: An Experience in Large-Scale Model Transfer and Evaluation, Part III by G. Johnson and J. Brown, Texas Energy and Natural Resources Advisory Council, Austin, 1980.

15. The definition of "reality" of course will differ depending on which philosophical stance one takes. Positivists do not believe values are real in the sense that physical objects distinguished by the five senses are "real." Normativists of course will not agree.

16. Massachusetts Institute of Technology, Energy Laboratory, Independent Assessment of Energy Policy Models: Two Case Studies, Rep. No. 78-001, Cambridge, Massachusetts, MIT Energy Lab., 1978.

C. Roger Glassey

10. Large-Scale Model Transplants from the Donor's Point of View

During 1979, the Energy Information Administration (EIA) successfully transplanted a very large, complicated modeling system, the Midterm Energy Forecasting System (MEFS) to the Texas National Energy Modeling Project (TNEMP). The process turned out to be a bit painful for the donor, but, unlike the case of heart transplants, this donor survived the operation. In discussing model transplants, I will mention a little philosophy, some law, a piece of history, and, finally, some observations about what we have learned from this experience.

First, the philosophical question--how scientific is energy modeling? One may regard energy modeling as essentially an art in which models facilitate computer assistance to the judgment of those forecasting or analyzing policy. A large model has several advantages. It allows integration of the knowledge and judgment of a large number of analysts, and it forces internal consistency on the resulting output. For example, a model can provide an accounting structure within which supplies and demands balance. A model can go further and bring to light inconsistency among the assumptions made with respect to exogenous parameters, such as economic growth rate and world oil prices.

There is another view, however, often espoused by modelers, which is that energy modeling is a branch of science. That, I suppose, is why the subject was discussed at the annual meeting of the American Association for the Advancement of Science. From this perspective, an energy model is an embodiment of a scientific theory. It is a representation of the way in which a part of the energy-economic system is believed to operate. There is a theoretical basis for the structure of the model, the definition of the major components, or variables in the model, and the way in which these variables interact. Furthermore, the coefficients of the model, demand elasticities for example, are estimated from observed data. This scientific view has certain important consequences. It implies, for example, that the theory underlying the model be explicitly stated. It requires that the method by which the coefficients were estimated and the underlying data upon which the model is based be accessible to inspection and replication by other workers in the field. Not only the model itself, but the way in which it is used to produce forecasts or analyses should similarly be made explicit, recorded in detail, and subject to peer review. The hope is, of course, that this process of peer review will detect deficiencies in the modeling system and in its application and that, as a result, both will be improved.

I should emphasize that at EIA and, I suspect, elsewhere in the modeling community, the model itself is only a part of the analytical process. It is only that part of the process which it happens to have captured, as a matter of convenience, in the form of a computer program, and it usually carries on the bulk of the computational work. It is by no means a representation of the entire analytical process and focusing on the model itself to the exclusion of the rest of the process is to miss a great deal of the important action.

Now for some law: The Energy Conservation and Production Act of 1976, P.L. 94-385, required the Federal Energy Administration (FEA), which was the precursor of EIA, to provide Congress and the general public with access to the PIES computer model on the FEA computer. PIES is the ancestor of MEFS; it is a very large, complex system of models, not a single model. It is composed of a number of models that independently represent energy supply, that is to say, coal and oil and gas, together with models that represent energy demand by end use by sector. Coupling these supply and demand models is an integrating model, which represents the action of the energy markets. The integrating model accepts supply and demand curves from the subsidiary models and computes the equilibrium supplies, demands, and prices. Imbedded within the integrating model are simplified representations of oil refineries and electric utilities.

The required access to the PIES model on the EIA computer has not been provided in exactly that form. Access to the model by members of the Congress, the Congressional Committees staffs, and branches of the Department of Energy has been provided indirectly. The Office of Applied Analysis within EIA has performed a number of analyses using PIES upon the request from these agencies. It does not, however, have the personnel resources or the computer time to do such analyses for people outside government. Access is provided not on the EIA computer, but by model transfer.

Now for some history: The National Energy Plan was published in April 1977. It has a strong demand management, or conservation, orientation. The report did not emphasize increasing supply. It contained one table, an energy balance table, which was produced with the aid of the PIES model. Critics of the National Energy Plan, particularly in the State of Texas, expressed some dissatisfaction about the numbers that were produced to support it. The Texas Energy Advisory Council (TEAC) requested a copy of the PIES model, a request which went to the White House Policy Office. This happened at about the time the Department of Energy was organized in the fall of 1977. The DOE Organization Act set up EIA separate from the Under Secretary for Policy and Evaluation, and EIA was divorced from any role in policy design and advocacy, whereas the predecessor organization under FEA had been involved both in analysis, design, and advocacy of policy. The request from TEAC ultimately arrived at EIA, which put together a collection of computer tapes representing the model and its data bases and sent them off. The history of this transfer was replete with fiascos, arising from such things as machine incompatibility and inadequate quality control in EIA (which once resulted in sending off a blank tape, much to the consternation of the people in Texas). Operating system incompatibility, less than adequate documentation of the models and how they were to be used, and a number of other minor difficulties contributed to our mutual problems.

The models involved were the oil and gas supply model, the coal model (the National Coal Model), RDFOR demand model, and the integrating model. Let me emphasize that this is a loosely coupled system of independent stand-alone models. It is not a single, simple model. Effectively orchestrating the operation of this collection of models by people who have had years of experience with it is not easy. Communicating the art of this orchestration to another group who did not have the benefit of experience turned out to be quite difficult.

In fact, the problems of reproducing the numbers of the National Energy Plan turned out to be formidable. Instead, the efforts of TEAC were redirected to focus on the 1977 Annual Report to the Congress. They did succeed finally in reproducing the Series C Projections of that document. Last summer (1978), a large conference was held where representatives of TEAC met with people from EIA to present the results of their study. During that conference, a number of defects in the modeling system which were known or suspected by EIA were confirmed by TEAC, and numerous suggestions, many constructive, were made for future development of the modeling system.

From the point of view of EIA, even though we learned quite a lot, it was a fairly expensive lesson. It cost about one to two top-level man-years as our contribution to the effort. We were a little surprised at how difficult it turned out to be, because model transfers are a fairly routine operation for us. Many of our models are developed by outside contractors and they are transferred from the contractor to EIA, where we routinely operate, update, and maintain them. However, the problems involved in transferring a collection of loosely coupled models turned out to be much more formidable, because the procedures for transferring data at the model interfaces were complicated and, as we discovered, not very well documented.

Our long-range goal now is to make our models portable and our data accessible to a wide public of experts. There are several reasons for wanting to do this. First of all, the ability of people outside EIA to reproduce our results enhances the credibility of the results. Second, as more people understand the nature and structure of our models, we have greater opportunities for learning how to improve them. Third, I believe it promotes rational debate on important energy issues if all parties to the debate have access to the same information. However, in order for these goals to be achieved, the receiving organization must make a substantial commitment of time and people in order to use the models effectively.

It may be useful to distinguish among several levels of understanding or usefulness of models and several kinds of users. The minimal level of understanding is to be able to carry out the mechanical tasks for feeding in the data files and the parameters settings in order to obtain results. A second level of understanding is required for intelligent use: that is the specification of parameters setting and the selection of data files in order to analyze a new question not previously posed to the model. A third level of understanding is required to modify the model code to simplify or enhance it, and thereby to expand the set of problems to which the model can be addressed. Users include first, the originators of the model who presumably know the most about it; second, the other technical people within the originating organization; third, technical people outside the originating organization; and fourth, nontechnical users.

The level of understanding required depends upon the category of users and their intended uses of the model. This level of understanding partly determines documentation requirements. In transferring the model within the originating organization, one can rely to some extent upon oral tradition as a substitute for documentation. When the model is to be transferred outside the organization, however, the requirements for completeness, formality, and currency of the documentation increase drastically. One of the costs of transfer is either to upgrade substantially the level of documentation beyond what is required by the effective functioning of the organization itself or to provide access to the oral tradition. This latter approach entails expensive long-distance phone calls.

Experience in transplanting PIES to TEAC has reinforced our determination to upgrade the level of documentation of our models. We have identified five classes of documentation, each of which is aimed at a different type of audience. First, model summaries outline the model structure, the theory on which it is based, and the extent and detail of coverage with respect to products, geography, time, process, and economic sector. The intended audience is the non-technical or casual user of the model; for example, someone in the Congress or in the policy branch of the Department of Energy who might use the results.

Second, the model description, which is an expansion of the model summary, contains much more technical detail at the level of a technical journal article. It is aimed at an audience of model builders and policy analysts. The model description document should also describe model validation results, what the aggregate properties of the model are, and what sorts of policy questions the model can suitably address.

The third category of documentation is the data description, which describes data sources and estimation procedures, and is aimed at the same audience as the second category. A fourth category is the users guide and operations manual for the technical person who will actually operate the model on a particular computer system. Finally, a fifth category of documentation is the annotated computer code and the details of the computer implementation for the technical person who intends to modify or possibly rebuild the model.

In addition to documentation, help is required to assist the receiver to adapt the modeling system (which has evolved within the EIA computer environment to take advantage of the peculiarities of that environment) to the user's software and hardware. There are very few computer systems in the country that are identical with respect to both hardware configurations and software operating environment. We face a serious tradeoff question, therefore, with respect to model portability. The cost of individual model transfers is reduced if the models are initially built to be largely independent of the EIA machine environment. This strategy, however, requires that we use common low-level languages and that we ignore certain features of our machine that are required for computational efficiency and efficient use as defined by EIA users.

Model portability requires not only much more detailed documentation than in-house use, but also a great deal more work concentrated on the ease of end use, in particular at the interfaces among pieces of the modeling system. Portability requires providing versions that will work in a wide variety of machine and software environments. Some estimates suggest that a commercial software product costs about three times as much to develop as a piece of software intended solely for in-house use.

Model portability benefits EIA through the ability to reproduce results, but there are limits, however, to how reproducible the results can be. For example, some of our models are solved by a linear programming system. Due to differences in the round-off conventions used in different computers, alternative solutions may result and, consequently, the values of the activity variables may be different. Differences in versions of compilers and in the linear programming codes that are used can also produce differences in the numerical values of the output.

Another benefit of access is that it promotes and encourages counter-modeling. If the participants in the energy policy debate have access to the same data bases and computer systems, they can then focus on disagreement about assumptions and theories of what is important in the geological, physical, or political world, rather than argue about the numerical values of the results. A third benefit is that model transplant provides an incentive to spruce up for inspection both the model and the accompanying documentation. In our experience with TEAC, we found how poor our documentation really was, and it provided us with an incentive to upgrade it. Finally, the outside review confirmed many of our suspicions about the areas in which our models needed to be improved and provided us with more confidence in setting the next phase of model development.

Let me conclude with the observation that the easier it is to transplant a model, the less interesting it is likely to be to the recipient. Several stages occur in the evolution of a model, and the rate of change of the model structure decreases during these stages of development. Consequently the possibilities of model transplant become easier. The initial stage of course is conceptual design when we do not have a computer program to transplant. The second stage is development, in which the model design is refined and captured in a computer program that is tested and modified. The third stage is routine use for analysis. In this the model continues to change. A new analytical question will typically require some modification. In some cases the modifications may be trivial, such as respecifying values for the parameters. In other cases, major structural changes may be required, such as incorporating a large, new segment in the Midterm Energy Market Model to represent the operations of the Natural Gas Policy Act, which we analyzed in the summer of 1978. Finally, when the latest version of the model is archived, the rate of change drops to zero and the transplant problems are minimized.

EIA policy now is to archive annually the versions of all models used to produce the Annual Report to Congress. These models will be available for public distribution through the National Technical Information System (NTIS). We instituted a quality control program to test the packages that are shipped to NTIS, to ensure that the descriptions for using the model are sufficiently complete. Thus a person who is not involved in active model operation should be able to follow the directions and actually produce the results of the test data that accompanies the tapes. These versions of the models will be about six months older than the ones that are currently used for analysis within the EIA. I hope that this archiving process will provide us with the benefits of model transplants: namely peer review and feedback concerning possible areas for improvement in model structure. As long as the EIA computer remains overloaded, however, there does not seem to be any possibility of providing public access to current versions of the model on that machine.

William W. Hogan

11. Alternatives to Transfer of Large-Scale Models

Introduction

The eruption of the energy crisis, part way through the decade of the seventies, created a fertile field for the use of large-scale policy models. Because energy problems are complex, energy decisions can affect every sector of our economy. With the abundance of data, analysts can build detailed models to describe these many effects. Rapidly changing energy markets unsettled everyone and generated a steady stream of new questions necessitating new models and new uses of models. As a result of this sudden surge in demand, energy modeling overnight became a growth industry. The Energy Information Administration (EIA), for example, reported over sixty models in use in the then newly formed Department of Energy (1).

This widespread use of large-scale models, compressed in time and visible in energy policy debates, raised anew important questions about the role of models in affecting policy decisions. The computer technology of the sixties provided the capability for large-scale modeling, promising improved decisions through computer analysis. But by the beginning of the seventies many of the promises were unfulfilled, the "whiz kids" and their computers had lost some of their sheen, and sophisticated critics began to ask: Were the models accessible? Were the models valid? Were the modelers fudging? How could the user check on the modeler? These were good questions for any use of policy models, questions that were ripe for attention when the boom in energy policy modeling began.

The importance of energy issues, the many applications of policy models, the concern of model users, and the spirit of the times combined produced a series of efforts to probe into the models and their uses. The new cadre of energy policy modelers was eager to refine the questions and develop the means for assessing policy models. Sponsors of model development recognized the need for model assessment and created new institutions such as the Energy Modeling Forum, the Utility Modeling Forum, the MIT Model Assessment Laboratory, or the Texas National Energy Modeling Project (2). The EIA created an Office of Validation (3). These experiments with policy model assessment should produce information applicable to all fields of policy analysis, not just energy policy analysis.

A common element in any evaluation of the accessibility and validity of a large-scale model is possible transfer of the model for use by someone other than the model developer or principal user. A transferable model is said to be portable, and portability is seen as an unambiguously attractive feature for any model. In the analogy to scientific experiment, evaluation of a theory, or a model, depends upon independent replication of results. The need for the transfer of large-scale models is self-evident, but the cost is high and the successes are rare. The difficulty of transferring any large model suggests that we be sure of the need before we undertake the task. A description of the need for portability may suggest alternatives that both lessen the difficulty and facilitate the use of the models.

In this paper I critique the need for, describe the obstacles facing, and summarize the major alternatives to transfer of large-scale policy models. The examples are from energy policy applications, but the principal arguments apply to policy models in general. I begin, in the next section, with a discussion of the concept of model validity and the reasons for transferring a model. A discussion follows first of the difficulties and costs associated with the transfer of a large model, and second, of alternatives to the complete transfer. The concluding section suggests a few implications of developing better capabilities in model transfer.

Model Validity

Transfer of a model begins with concern over the reliability of the model. If a model is to be useful for policymakers, they must have confidence in its accuracy. Assumptions should be justified and clearly stated, calculations must be properly done, model structure must accurately describe the system modeled, and results must be presented so as to be easily understood and used. If these conditions are not met, at best, the decision-maker will be abdicating his responsibilities in favor of the modeler; at worst, the decision-maker will make bad decisions based on false or confusing information. An essential part of modeling, therefore, is knowing when the model is reliable, and how to use it within the limits of its reliability.

This characterization--the limits of reliability--implies that there are degrees of acceptance of a model. Evaluation of any model depends on the circumstances and the model's intended use. And model assessment has different phases, each with its own goals, tests, and standards. Viewed from the perspective of differences in goals, three distinct phases emerge: ventilation, verification, and validation (4).

Ventilation is the examination of the assumptions, theory, and structure of the model. This is the primitive phase of describing what the model does and how it does it; in many contexts, this step is so trivial as to be taken for granted. In the analogy to theories in the physical sciences, for example, there is seldom any difficulty in describing the model. A few equations suffice and the relationships are transparent. There may be profound difficulties in designing and conducting tests of the theory, but the model of the apple falling from the tree is plain for all to see. Not so with large-scale policy models, which involve many equations with many interconnections. The link between environmental regulations and the demand for coal, for instance, can be complicated by regional variations in standards, treatment of old plants, and pricing of competing products such as natural gas. A formidable task confronts the user in understanding what the modeler claims to have built; as with any large system, feedback can surprise even the modeler. The user and the modeler must ventilate large-scale models to expose the structure to scrutiny, to clarify the critical judgments, and to lift the curtain between the modeler and the user.

Verification is the examination of a model to ensure that the modeler has faithfully implemented the details as designed. There are many opportunities for error in the description of the equations, the estimation of the parameters, and the solution of the model. A decimal point misplaced, a variable dropped, or an assumption violated may be difficult to detect and yet may significantly alter the results. Every change in the model provides an opportunity for creating, and finding, errors. The modeler must test constantly to verify that the model is as intended. With a large-scale model, the modeler bears the burden of proof that the results flow from the theory, data, and assumptions, not from a mistake in implementation.

Validation is the demonstration that the model is a faithful representation of reality. This is the ultimate test. If the model is valid, then the results can be believed and used. In the analogy to the testing of scientific theories, we evaluate the model through comparison of the predictions and experimental evidence. If the model predicts well in the experiments, then its predictions can be accepted for other cases and other assumptions. This suggests that validation is a process, and no model or theory is ever finally validated through experimental evidence. If the model predicts incorrectly, it can be rejected as invalid or, at least, modified and improved; but the modeler can construct ever more demanding tests to validate the model. The greater the variety and the more difficult the tests that the model passes, the more our confidence in the model grows, and the more we begin to describe the model as valid, reliable, and correct.

With a validated model, verification is implicit and there is less need for ventilation; we are more willing to treat the model as a black box, accepting the results without always understanding every step from assumption to conclusion. But this is asking a lot, because for many reasons it may be difficult or impossible to obtain validated policy models. Policy questions, for example, are often pressing and contrafactual. The decision-maker wants to know how a system of price regulation will affect the dynamics of supply and demand. Different incentives will be created by the regulations, and many variants of the regulations are being considered. There is no record of past data to draw on; seldom is there either the time or the inclination to conduct experiments with each new version of the regulations. The decision-maker needs a model to evaluate alternative proposals, but there is little hope of validating more than a few components of any model.

The deadline pressure of policy decisions may preclude even the validation of apparently scientific models of physical processes. Policy decisions surrounding the disposal of nuclear wastes are a case in point. It is inevitable that disposal decisions will be made soon on the basis of predictions of models concerning the behavior of highly concentrated, high-level wastes stored for thousands of years in a variety of possible depositories. In this case, we are likely to rely more on judicial procedure than on scientific method in marshaling the facts to assess and evaluate the models.

If experimental validation is beyond the reach of policy models, then judgment, based on partial information, becomes critical in evaluating a policy model and in deciding how much the model should influence the decision. For policy models, therefore, ventilation is the essential phase in evaluation of a model. The technical details of verification are usually of little interest to the policymaker, who is content to let the modeler check the arithmetic. But the judgments, the critical questions that cannot be answered with experimental evidence, are more important. The policymaker will perceive the critical judgments as his or her prerogative and responsibility. To make these judgments, the decision-maker must understand the essence of the model and the implications of alternative assumptions. Often this is all the policymaker wants or expects--to understand how the model works. This need, to ventilate the models, is behind the development of the modeling forums, where modelers and users compare and contrast alternative modeling approaches to the same policy problem (5).

Model Transfer Demands

Variations in the goals and means of model assessment and use produce a similar variation in the goals and requirements for the transfer of large-scale policy models. Complete model assessment and independent replication of results, by definition, require the transfer of ideas, data, procedures, and software. For the ultimate scrutiny and test, there is no substitute for the ultimate transfer of everything associated with the model. Many uses of a model do not require full transfer, however, and alternative approaches to model access provide the benefits of portability without some of the costs of a full transfer.

Ventilation of a model does not depend upon the transfer of the model to a third party. In the most familiar approach, written descriptions and documentation communicate the essential structure of a model. Written well, a paper can isolate the critical assumptions and proper uses of a model. In many important instances, of course, the modeler fails to anticipate the questions, and the documentation is not useful, or the user cannot penetrate the technical jargon to find the vein of key ideas. Ventilation of the model requires a bridge between the technical detail of the modelers and the broad policy needs of the decision-makers. We build such bridges through the close interaction of the modeler and the decision-maker working toward a common understanding of the problem and the role of the model. The closer the model is to the decision-maker, and the more likely the interaction, the greater the use of the model (6). This interaction is the core of the Energy Modeling Forum (7).

A good model will be in demand. On the surface, it seems that the best and easiest way to use a model is to obtain a copy of it. Hence, in many cases, the demand for analysis will create a demand for model transfer. But this is not always true, and full transfer is not necessary for analysis. The critical need is to change the assumptions and rerun the model to apply the results to new problems. If we have confidence in the model and access to the modeler, we can obtain sensitivity runs without direct communication with the model.

Even for counter-analysis, that is, studies done by one decision-maker or modeler to oppose the results of another, we may be satisfied with indirect access to a model. The Energy Information Administration, for instance, can and does exercise its models at the request of policymakers, in the Congress and in the Department of Energy, with sharply conflicting views on the merits of the policies under study (8). In the best examples, both sides have implicitly accepted the same model, and the debate has moved beyond the description of alternative policies to the value judgments for the tradeoffs the policies require. Of course, the issues may be sensitive enough to prompt the user to seek access to the model without the filter of the model developer. But access is the key, and full transfer of the model is only one of many ways to obtain access.

The occasional transfer of a model may be valuable in creating the proper incentives and standards for modelers. Any model should be transferable, in principle, and it should improve the credibility of both modelers and models whenever a successful transfer occurs. Despite all its difficulties and all the problems it uncovered in the transfer of the PIES model of the EIA, for example, the Texas National Energy Modeling Project (TNEMP) proved that the model existed and an independent group could, at considerable expense, reproduce the results generated within the Department of Energy (9). This audit function, however, can be performed without requiring full transfer of the model.

Full transfer of a model, therefore, may serve many ends, but there may be many paths to these ends. The independent replication of results will guard against error and deception. For this function alone, full transfer may be indispensible. However, for other functions--ventilation, analysis, counter-analysis, and audit--full transfer may be convenient, but is not essential. Our concern with full transfer, therefore, can be tempered by the knowledge that it is needed only on occasion, and we can consider the costs of evaluating alternative approaches to achieving the same end.

Model Transfer Obstacles

Large-scale modeling is expensive. Development of the theory, identification and assembly of the data, design and operation of software, and translation and interpretation of the results are all demanding tasks that consume valuable resources. For the most part, however, the preparation of one more run of the model is inexpensive; there are large economies of scale in large-scale modeling for any purpose, and this is no less so for policy modeling. It follows, therefore, that there are natural economic incentives to centralize the construction and operation of a large policy model. Furthermore, the state-of-the-art of modeling is so primitive that the cost of transferring and testing a model can approach the cost of original development (10). Policy models are everchanging and, as a result, they are seldom mature enough to lend themselves to easy or straightforward operation. The operator must understand the model and how to detect and correct errors at the sensitive interfaces of the model's components; this requires a high level of skill and a large investment in education on the details of the model. As in the sciences, we may judge the value of independent replication to outweigh the costs, but the high cost will remain for a long time as an obstacle to full transfer of large-scale policy models as we know them today.

A good policy model has economic value. The interests of analysts in the transfer of ideas and the validation of theories may not coincide with the economic interest of the owner of the model. Many large-scale models useful for policy analysis are proprietary, in law or in fact. The developer of a model has incentives to resist the full transfer of the model and the attendant loss of economic control. There are solutions to this problem, of course, including fees and copyright protections, but these add to the cost of transfer and complicate the process.

Knowledge is power. This maxim applies especially to the world of policy analysis, which works entirely with ideas. Control over the analyses and the process of analysis, through control of the models, can lead to subtle but important influences on the decision-making process. By controlling the model, one actor may define the terms of the debate and the form of the answers. (This problem, of course, is not unique to modeling. It is the source of the growth in the staffs at both ends of Pennsylvania Avenue, where more and more competing analysts study the expanding list of policy proposals on a dazzling range of issues.) The monopoly control of analysis is as much an incentive for demanding model transfer as it is an obstacle to achieving model transfer. But it is an obstacle nonetheless; even with a good and valid model, the model developer may be disinclined to dilute the power that comes as a fruit of a successful model development.

The expense, the proprietary restrictions, and the dilution of power can or should be overcome. If we take the transfer seriously, we can hurdle these obstacles by applying sufficient resources or by developing the appropriate institutions. However, other more subtle problems present more serious challenges to the ideal of transferring large-scale policy models.

Policy debates change and the models change apace. The more important the policy problem, the more complicated the analysis, the more controversial the results, and, therefore, the more important the independent assessment of the models, then the more likely that the models will be evolving as the modelers incorporate new insights and adapt to new perceptions of the policy problems. In the heat of battle, there may be little time for model transfer; the modeler is more likely to be struggling to insert the latest modification into the model. Anyone familiar with real-time policy analysis knows the tension created during the short period when the need for consistency leads to the suppression of model improvements or even the accommodation of small errors. Because of the tension, consistency does not hold sway for long; new components replace old, and the model marches through its many generations. Hence, by the time the modelers can complete a full transfer, the recipient has an outdated model and the policy debate has gone off in a new direction. The developer, in responding to critiques of the model, correctly reminds everyone that the comments bear little relationship to the existing model in its current uses.

This problem is endemic to the policy process. The window for policy decisions often is barely adequate to develop a good model, much less to transfer the model and subject it to extensive testing. It is not likely that the policy process will wait for the models. Foregoing the models because of inadequate evaluation just leaves the decision-maker with only judgment and expert advice, decision aids which may be less effective without the models and which are certainly less explicit and less open to evaluation than the models. The solution, therefore, must be found in the development of alternatives to the full transfer and independent assessment of the models.

In a related problem, the complexity of a policy problem and the size of a large-scale model often create an essential dependence of the model on the modeler. A large-scale model is a complicated instrument and, as with any instrument, it performs better in some hands than in others. Few policy problems are so well understood that the model incorporates everything of importance. Changes in assumptions and adjustments of parameters may allow a model to adapt to a new problem. But there is as much art as there is science in this process, maybe more, and some artists have more talent and greater vision. It is part of the conventional wisdom, for example, that the "addfactors" in the major macroeconomic models, the entry points for the modelers' judgment, are as important as the models themselves. Eckstein and Data Resources, Klein and Wharton, Evans and Chase; the names are linked to the models and the models would not be the same in other hands. We prefer the model to be portable, but not necessarily at the loss of the expertise of the model developer. A complete transfer and independent evaluation of the model may be valuable for improving the models, but it may not be the best way to access and use a model, even if it could be done easily.

We see, then, that there are obstacles to the transfer of large-scale policy models, and portability may not be as valuable or as essential a characteristic as appears on first examination. We need a more selective approach to the problems that transfer is intended to meet. We need practical alternatives that gain some of the benefits of portability while avoiding many or all of the costs of full transfer.

Approaching Portability

There are many steps to take in approaching full-fledged portability of a model. Seven examples follow, in roughly increasing order of cost and approximation of full model transfer.

Contract Analysis

Most models and modelers are highly accessible, at a price; remember that large-scale modeling is expensive. Most modelers are willing, even eager, to discuss the intricacies, both strengths and weaknesses, of their models; but here the cost is different. The user must contribute time, attention, and effort to the development of understanding, to the construction of questions that probe the limits of the model. To be sure, this is not an easy process and it will not uncover everything that would be found in an independent assessment, but a sophisticated user can learn a great deal about a model through the direct interaction with the model developer. A little investigation can settle most of the issues that might lead a user to look for an independent evaluation of the model.

Access is a large part of the attraction of portability, and access is usually there for the asking. Given the economies of scale in a large modeling effort, it may be easier and more efficient to achieve access through a direct working relationship with the model developers. This is the most natural step to understanding a model. The evidence is that the most used models are those closest to the decision-maker.

Modeling Forums

The efforts of the modeler and the user examining a single model reinforce each other in the comparative studies of modeling forums. Forum working groups, composed of model developers and model users, ventilate the models by comparing the models in the analysis of a common policy problem. Focusing on a policy problem provides the interest for users and the concrete detail for discussion with the modelers. The contrast of modeling approaches suggests questions for all models, more questions than would arise naturally in the examination of a single model. Differences in accounting conventions, for example, which often explain differences in forecasts, will elude us in the examination of a single model, but become apparent in a comparison across models.

In the experience of the Energy Modeling Forum, the modelers help the users articulate their policy questions in terms compatible with the models and, in the process, help the users understand the limits of the models and their uses. This was expected and designed into the forum operations. What has been surprising, however, is the education the modelers receive from each other. In comparing their models, the modelers recognize and adopt new design features so frequently and so rapidly that models tend to converge during a typical forum study. On occasion, the modelers discover unknown limitations in their own models (11). These latter features of the forum studies capture some of the benefits of an independent assessment, without the cost of a model transfer. There is, of course, no protection against intentional deception, but for efforts of good faith, the forum studies provide a check on errors, omissions, and self-deceptions.

Remote Access

The DRI macroeconomic model is a sophisticated, large-scale model of the American economy (12). Expert modelers maintain and operate the model, constantly updating it to incorporate new information and meet new needs. Furthermore, this model is, in a real sense, highly portable. For a fee, this model, with all its data, equations, parameters, and supporting software is available to me, as close as the nearest telephone. The model resides on a central computer operated by DRI and accessible through a terminal connected to a telephone network. In principle, I can conduct any test and make any change, just as I would be able to if the model were sent to me on magnetic tape and I installed it on my own computer.

The advantage, of course, is that I can use the model and conduct limited tests without absorbing the costs of a full transfer. If I choose, I can concentrate on a few features of the model and rely on others to treat segments of less concern to me. Hence, I might change the equations in the energy sector without steeping myself in the arcane details of the characterization of the financial sector. In addition, I can always have access to the latest version of the model, with the DRI modelers' best judgments on the proper assumptions for the long list of input parameters.

In fact, remote access is almost a perfect substitute for the full transfer of this model. Barring the improbable event of a massive fraud orchestrated by DRI, the only added benefits of a complete model transfer would be (1) the possible discovery of a computer software error, which, to have survived this long, must be both obscure and subtle; and (2) being forced to examine and understand all, not just part, of the model. There are cheaper ways to check on computer errors; we could develop other incentives for self-discipline.

Not all models are as suitable as the DRI model for remote access. There are few models, however, that we could not adapt to this mode of operation. With the continuing decrease in computing costs and the expansion of computer networks, we should look to this example and this approach as possibly the best compromise for achieving the benefits of portability and full transfer of large-scale policy models.

Models of Models

Part of the difficulty of transferring a large-scale model stems from its size and complexity. If we could make the model smaller, we could make it portable and easy to use. If we could simplify the model, we could understand its properties and predict how it would perform in circumstances that satisfy the assumptions of our simplifications. In short, if we could build a model of the model, we could reduce the cost of ventilation, analysis, and use. Just as we built the large-scale model to study the larger and more complex world, so too can we build a small-scale model to study the large-scale model.

This approach to studying models is more than just a matter of convenience. For many purposes, the simple model may suffice for our analysis but we cannot estimate the model from the noisy real-world data available. Consider, for example, an aggregate model of the energy sector with a single parameter for the elasticity of energy demand. If all energy prices were increased proportionately, this model would be a legitimate aggregation of the energy sector and could give good predictions of the response of energy demand. The real-world data, however, never satisfied this aggregation condition, and it can be shown that almost any aggregate elasticity could result, depending upon the mix of price changes (13). If we have a large-scale model of energy demand, however, we can conduct a controlled experiment with proportional price changes and estimate the demand elasticity consistent with the aggregation condition. The resulting simple model would be portable and easy to use, and it would yield insights into the structure of energy demand, insights that could not be obtained directly from real-world data.

Models of models are, in fact, pervasive in large-scale modeling. Modelers usually build a large model by combining smaller models of components of the system. An econometric model will be used to describe energy demand, a detailed process model will be used to describe energy supply and conversion activities, an input-output model will integrate the energy sector into the rest of the economy, and so forth. Often the design, accounting conventions, and level of detail in these component models are incompatible. The modeler links the components by building an interface--reduced form models for demand, extreme point models for refineries, and so on--and connects these models of the models to create the large-scale system (14). This approach to model construction has many names: pseudo-data, response surface, approximation, reduced form, cartoons. However, modelers have only begun to tap this approach to model assessment and access (15). It may be, someday, that every large-scale model will have an associated set of small models available as special-purpose simplifications to promote access and portability. This would be the logical

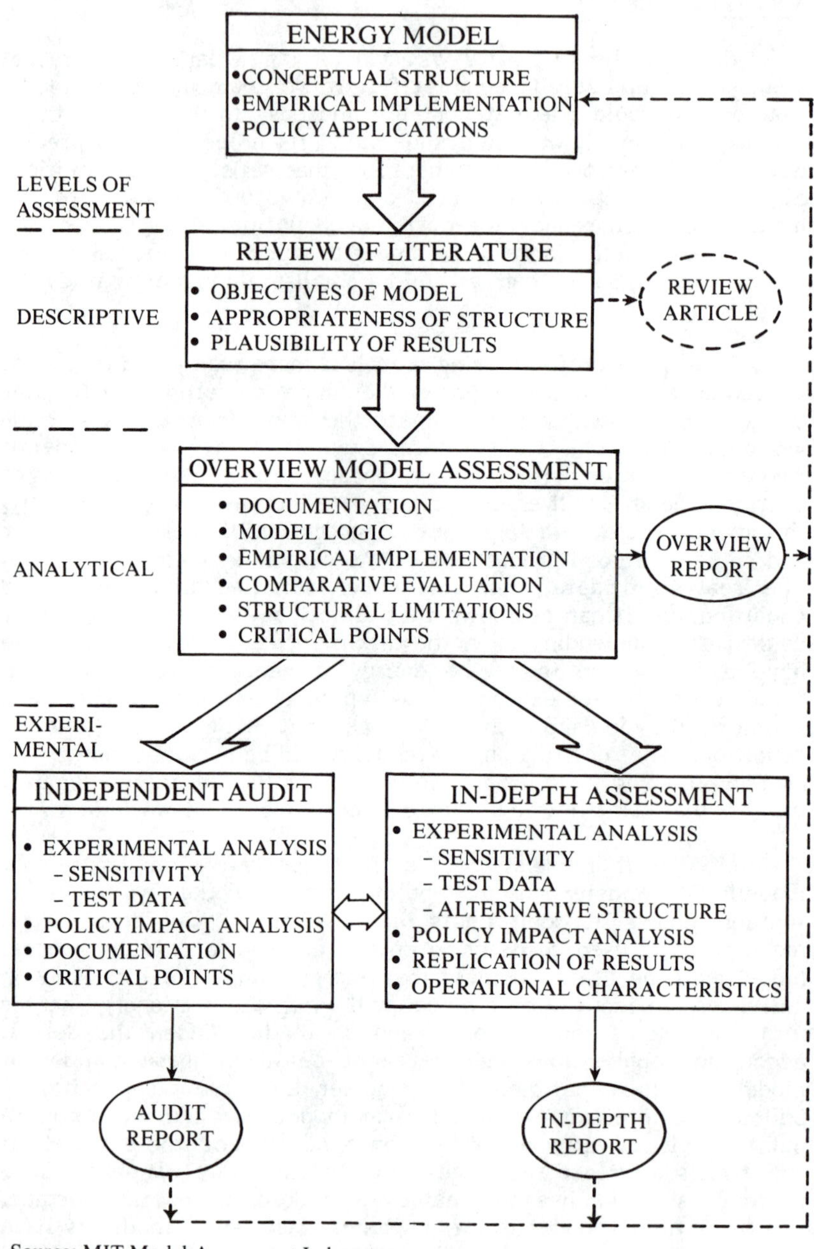

Source: MIT Model Assessment Laboratory.

Figure 1. Approaches to Energy Model Assessment.

extension beyond the creation of point forecasts and sensitivity analyses.

Model Audits

The MIT Model Assessment Laboratory has been developing and implementing an approach to model assessment. Motivated by the scientific analogy, the analysts have been attacking all three phases of assessment: ventilation, verification, and validation. In a taxonomy orthogonal to these phases, they describe an approach to model assessment that includes many steps or levels of analysis (see Figure 1). Each level of assessment reveals more about the model, and we can learn a great deal before we reach the most expensive stage, the in-depth assessment. Yet it is only at this stage that we require true independence from the model developer for independent replication of the results and, therefore, it is only at this stage that we need complete portability of the model. Even the independent audit can be conducted through remote access.

If our purpose is to subject the model to outside review and demanding tests, then the MIT work suggests that full transfer may not be necessary in every case. The central requirement, again, is access, and access can be obtained without absorbing the cost of full transfer. At the same time, we see that the audit of a model, through the several steps of a model assessment, is expensive. It requires time, resources, and talent on a scale similar to that of the model development. A sign of maturity in policy-modeling will be the widespread recognition of the need for careful model audits and the attendant willingness to pay the price.

Standard Software

Modelers assembled the earliest large-scale energy policy models with the modeling counterparts of bailing wire and chewing gum. The models were as rickety in flight as the first airplanes. *Ad hoc* adjustments, human intervention in the running of the model, and laborious hand calculations at the interfaces characterized the operation of large-scale models. With more experience, however, the modelers recognized a common structure in many of the components and began to move toward similar software for building and manipulating a large model (16).

Improved software offers two main advantages. First, it simplifies the construction of a model, and simplified construction creates the capability for the second advantage, an improved portability. With flexible software, the reconstruction of a model becomes inexpensive; transfer no longer requires the exchange of computer codes and the management of the intricate interfaces between model components. Transfer of a model becomes the transfer of data and ideas. Given the data, and understanding the ideas, the recipient modeler constructs the model and, in the process, provides the important test of the true independent replication of the system.

This procedure will, in many ways, be preferable to the actual transfer of the original software. The recipient modeler will gain an intimate understanding of the model structure and, therefore, be more likely to discover the soft spots in the design and the data. If, in addition to assessment, the purpose of the transfer is to modify the model and use it for further analysis, the reconstruction may be cheaper, in the long run, than the straight transfer. Every model has its idiosyncracies, developed over time as the modeler changed the system while moving from one problem to the next. Every modeler has ideas about how his model could be improved if only he could start over again, but the cost is too great. During a model transfer, however, the cost must be incurred and the improvements can be had almost for free. The best way to transfer a model may be to study it carefully and then take the ideas, leaving the corpus behind. With a flexible, standardized software for policy modeling, software of the type now under development, true portability may come naturally as part of the practice of modeling.

Model Transfer

In the end, for the narrow purpose of independent replication of results on specific policy studies, there is no alternative to a straight and complete transfer of a model. There are cases where the importance of the problem and the sensitive, contentious nature of the issues demand the test that can be performed only with a complete model transfer. The TNEMP examination of the PIES model and its uses in the formulation of the National Energy Plan may be an example of a case of sufficient gravity. Before undertaking such a large effort, however, we must recognize the large cost and the small incremental benefits. For almost all purposes other than independent replication to test for undetected error or to guard against deception, there are attractive alternatives to complete transfer of a large-scale model. The high premium we place on portability of models may be too simplistic. Through a deeper examination of the concerns with any model, we can better match ends and means.

Conclusions

Portability is an attractive feature for a model. If the transfer of a large-scale model were inexpensive, quick, and effective, then transfer would be the easy solution to a wide range of problems associated with understanding and using policy models. Unfortunately, the transfer of a large-scale policy model can be expensive, slow, and, most importantly, may compromise important features of a policy model--the currency of the formulation and the access to the insight of the modelers. Given the many obstacles to the full transfer of such models, we should search for alternatives that provide the same benefits as portability but avoid the costs of a full treatment.

Many alternatives are available. If the need is for access, then direct interaction with the model developer can be an effective means for using a model. If the need is for ventilation, the most important phase of a model assessment, then modeling forums present an attractive alternative. If the need extends to using the model as part of a larger system, then it may be easier to construct a model of the model than to perform a full transfer. Even for a model audit, there are many steps to take before a full transfer of the model becomes indispensable.

Nearly complete portability can come through remote access. Development of remote access capability for most large-scale models would be a less drastic step than periodic transfer, and it would be more compatible with the objectives of model developers in the construction and use of large-scale policy models. Remote access has met the demanding standard of the competitive marketplace, but modelers and model users have not given it enough attention as a mechanism for meeting the different standards of model assessment.

If transfer seems essential, all but the narrowest purposes could be met through the efficient reconstruction of the model. Development of better modeling software, a trend supported by the needs for easier model construction, will make this approach to model transfer more prevalent in the future. The implication is a change in the focus of documentation, with an increased emphasis on the full disclosure of data bases.

Finally, we must recognize the cost of model transfer as part of the normal costs of policy-modeling. The trend is there, and it should be reinforced. Sponsors should continue to expand their requirements and budgets for model documentation, access, and assessment. Modelers must consider these phases during the process of model development. Users must participate in the process of assessing and using models. Policy problems are too important to delegate to a priesthood of modelers; hence, there is no danger that any such delegation will last too long. To avoid the equally extreme action of rejecting large-scale policy models as arcane and unreliable, everyone with an interest in better modeling must develop new ways to look at models, alternate approaches to their ventilation, standards that derive from the needs of policymakers, and a process wherein users are the masters of models they exploit.

Footnotes

1. Energy Information Administration, Annual Report to Congress, Volume 1, Department of Energy, 1977.

   This report lists sixty models in place in 1977. The EIA report, Models of the Energy Information Administration, May 1978, lists sixty-three models.

2. The origins of the Energy Modeling Forum are described in Hogan, W., "The Energy Modeling Forum: A Communication Bridge," in Operational Research '78, K. Haley (ed.), North Holland Co., New York, 1979: Also in the Sweeney, J., "The Energy Modeling Forum: Improving the Usefulness of Models," presented at this symposium. The Utility Modeling Forum began in 1979 and operates under the direction of Robert Shaw of Booz-Allen Inc., Washington, D.C. The Texas National Energy Modeling Project, operated under the direction of Milton Holloway, is described in Holloway, M.L., et al., Texas National Energy Modeling Project, Texas Energy Advisory Council, Austin, Texas, 1979.

3. The responsibility of the office, known formally as the Office of Analysis Oversight and Access, Energy Information Administration, extends beyond the function of model validation. The director is Dr. George Lady.

4. I suggested the term "ventilation" in the paper "Energy Models: Building Understanding for Better Use," Lawrence Symposium Proceedings, University of California, October 3-4, 1978. An orthogonal classification of the phases of model assessment can be found in Kuh, E., and D. Wood, "Independent Assessment of Energy Policy Models," MIT Model Assessment Laboratory Report to the Electric Power Research Institute, EPRI EA-1071, May 1979.

5. For a description of the operations of the EMF, see Sweeney's paper in this volume.

6. Fromm, G., W.L. Hamilton, and D.E. Hamilton, "Federally Supported Mathematical Models: Survey and Analysis," Report to the National Science Foundation, Washington, D.C., June 1974. See also Greenberger, M., M. Crenson, and B.L. Crissey, Models in the Policy Process: Public Decision Making in the Computer Era, Russell Sage, New York, 1976.

7. See the Sweeney paper in this volume.

8. The EIA involvement in the analysis of the Natural Gas Pricing Act is a case in point.

9. See the report on the Texas National Energy Modeling Project, 1979.

10. See the report by Kuh and Wood (1979).

11. The results of the first three EMF studies are reported in Energy and the Economy, EMF 1, Volume 1, Stanford University, 1977; Coal in Transaction: 1980-2000, EMF 2, Volume 1, Stanford University, 1978; Electric Utility Load Forecasting: Probing the Issues with Models, EMF 3, Volume 1, Stanford University, 1979. These reports contain many examples of the education of the modelers as well as the model users.

12. Data Resources, Inc., Lexington, MA, 02173.

13. See the paper by James Sweeney in the report of the Energy Modeling Forum, Aggregate Demand Elasticities, EMF 4, Volume 1, Stanford University, 1980.

14. For an overview of combined models, see Hoffman, K.C., and D.W. Jorgensen, "Economic and Technological Models for Evaluation of Energy Policy," The Bell Journal of Economics, Volume 8, Number 2, 1977. The PIES model is a major example of a combined model; see Hogan, W., J.L. Sweeney, and M. Wagner, "Energy Policy Models in the National Energy Outlook," in Energy Policy, J.S. Aronofsky et al. (eds.), Studies in Management Sciences, North Holland Company, 1978.

15. See the first report of the Energy Modeling Forum, Energy and the Economy, EMF 1, 1977, for an example of the use of a small-scale model to describe the behavior of large-scale models.

16. The SRI-Gulf model and its descendants have been the object of serious efforts to develop standardized modeling software. One effort is centered at the Lawrence Livermore Laboratory at Berkeley. A second effort is centered at Decision Focus Inc. in Palo Alto, California. See Sussman, S.S., and W.F. Rousseau, "A Demonstration of the Capabilities of the Livermore Energy Policy Model," Lawrence Livermore Laboratory, July 1, 1978, for a description of the former effort, including the use of microfiche for a documentation of the model structure and results. See E.G. Cazalet, et al., The DFI Computer Modeling Software, DFI report to the Energy Information Administration, December 1978, for a detailed discussion of the DFI approach.

James L. Sweeney

12. Energy Model Comparison: An Overview

Abstract

Organization and manipulation of complex data bases always requires models of some sort, and all models, being of necessity simplifications, are imperfect. Systematic model comparisons can produce benefits through identification of errors, clarification of disagreements, and guidance for model selection. Model comparison categories include methods and equations, forecast, aggregate behavior, and model regeneration. Modeling-the-model helps to structure and to communicate these comparison results.

Introduction

The policymaker or planner confronting any system as complex and controversial as the energy system must assimilate and analyze a vast array of information about technology, existing policies, regulations, economics, and political forces, and anticipate the reactions of constituents and other interested parties to any action. The planning and using of the requisite data base always requires models of some form--explicit or implicit, mathematical or intuitive--since a model is simply an organized set of cause-and-effect relationships and basic data. Although formal mathematical models may not be utilized, information about possible futures or about the effects of alternative actions on these futures must at least implicitly be shaped through modeling, unless the policymaker or planner depends primarily upon soothsayers for advice.

In organizing and manipulating information, judgments must be made about the relative importance of and relationships among diverse pieces of information. The basic cause-and-effect relationships accepted by the individual are shaped by previous experiences, judgments, evidence, and professional intuition. Therefore, the model developed by one person to describe a complex system will normally differ, often in important ways, from that developed by another person. Such model differences may lead naturally to conflicting conclusions as to what policy choices should be accepted, even when there is a common goal.

Whenever differences in policy or planning recommendations stem from discrepancies in the information base, an important aspect of debate resolution involves a comparison of the implicit and explicit assumptions made by participants in the debate. A system of examining assumptions is valuable for understanding the fundamental points of agreement and disagreement among the parties and for striving toward consensus.

While comparison of assumptions is crucial for achieving mutual understanding and consensus, the examination of information sources is often hampered by obstacles common to all applications of modeling, particularly formal modeling (7).

(1) Communication between developers and users of formal models is extremely limited.

(2) Formal and mental models often remain as black boxes, opaque to potential users, and translucent only to their developers.

(3) Although the various models may exhibit widely differing behavior, and therefore provide different answers to any question, they are seldom used in a comparative mode to address policy or planning questions.

A formal model comparison can alleviate many of the difficulties that are faced by modelers and model users. Development of a formal methodology of model comparison and assessment will complement the usual procedure of model assessment that naturally occurs with the development of a model. Formal assessments of a model focus on the evaluation of the quality of a model, the accuracy of its projections, or similarly normative concepts. Assessments imply a comparison between a structural model and a mental model or between two formal models. Although assessment implies comparison, the converse is not necessarily true: comparisons can be made without necessarily reaching conclusions about the relative quality of models.

I will next discuss the process of model comparison, without addressing the more difficult process of model assessment once the comparisons are accomplished. Benefits of formal, extensive comparison efforts are discussed, followed by an outline of some comparison methods. Finally I present a purely subjective view of the relations between the methods utilized and the benefits obtained by model comparisons. While formal model comparisons and assessments are being conducted intensively through groups such as the MIT Model Assessment Laboratory, the Utility Modeling Forum, and the Texas Energy Advisory Council, most examples will be drawn from the author's experience with the Energy Modeling Forum (12).

Reasons for Model Comparison

The essence of modeling is simplification. From the wealth of information potentially available, relatively few crucial cause-and-effect relationships are selected. These relationships form the basis of the model. Relationships and data that are judged less important are given little or no role in the modeling process. This simplification permits an emphasis on important issues and allows easy manipulation of a complex system. Thus the simplification process relies on individual human judgment. As a result, there are at least as many models as "modelers." Even if all modelers were to have the same basic perceptions of the systems being examined, they would still invariably develop different models, based upon their different time constraints, goals, styles, research budgets, organizational talents, motivations, and judgment.

In addition, basic perceptions and assumptions vary: probably no two modelers possess the same information or judgments about the relative importance of various relationships in the system. Since a model is a product of human understanding, limits to that understanding are translated into limits in our ability to represent the inherent complexities of the world accurately. What we understand imperfectly, we model imperfectly.

Because of this dilemma, modelers should examine the implications of each simplification. This examination inherently involves a comparison of models. At the earliest stages of research, the comparisons are normally made among several alternative conceptual structures formulated only into mental models. At later stages, comparisons are made between the evolving formal model and the results of mental models. Such comparisons are considered by most modelers as a necessary part of the model development process. At more mature stages of model development, systematic comparisons among the various formal models and mental models can further enhance understanding of the specific modeling simplifications.

Thus, the issue is not whether model comparison and assessment is important. Model comparisons of some form are conducted whenever formal models are developed or utilized and whenever people attempt to resolve differences of opinion. The resolution of those differences or the choice of one answer over another at least implicitly includes a comparative assessment of the several sources of information. The dilemma faced by modelers is the question of the depth of the model comparison. Model comparison is costly and time consuming. Resources devoted to comparison exercises may imply fewer resources available for model development or application. The appropriate mix of activities is not obvious.

Benefits of a formal model comparison are the identification of errors, clarification of disagreements, and guidance for model selection. These three areas of model improvement will be explored in turn to delineate the benefits that are possible with a formal model comparison.

Error Identification

Every modeler makes errors in the process of developing a model. The goal is to identify and correct those errors as early as possible. Some are identified when the formal model is first compared with a mental model. Some errors are not apparent by comparison with mental models but may persist until more quantitative comparisons with other formal models are conducted.

The simplest type of mistake, computer coding error, often can be identified by comparison between the evolving formal model and a mental model. However, some more subtle coding errors may persist until formal model comparisons have been conducted.

Conceptual errors may be more difficult to identify without formal comparison exercises. A common conceptual error is to calibrate a model based upon data which bear the right label but which measure the wrong concept. For example, oil resources can loosely be identified as recoverable resources or as oil-in-place. The three-to-one difference can raise havoc unless one knows which concept is being employed. Or, more subtle differences can occur between data on production of crude oil versus all petroleum liquids. Errors of this type have been found in completed models through the Energy Modeling Forum (EMF) studies.

Another example of a conceptual error is provided by the experience of EMF 2, "Coal in Transition" (2). This study analyzed several models of coal production and distribution that used optimization algorithms to simulate a competitive equilibrium. In one model, the optimization was based implicitly on the assumption that producers could charge different prices to different customers for the same commodity. Comparison of results from that model with results of other models made the error apparent--the model was implicitly simulating a monopoly solution when the competitive solution was desired. That observation led to restructuring the solution algorithm so as to modify the implicit behavioral assumption of the model to that of a competitive equilibrium system.

Clarification of Disagreements

The second major benefit of formal model comparisons is the clarification of points of contention and disagreement. This advantage itself serves as a major stimulus for the growing professional interest in model comparison and assessment. When inputs of the various models are standardized, the results and projections may still vary significantly. These remaining variations may motivate a search for differences in modeling assumptions, data, or parameters--differences which may be fundamental to policy conclusions. Underlying differences--"contention points" (13, 7)--can be examined, debated, and researched. In fortunate situations this process may lead to resolution of the differences. In any event, the isolation of the basic disagreement tends to focus and sharpen the debate and may lead to enhanced understanding of the fundamental issues.

For example, differing assumptions about how OPEC nations set production quantities can lead to differences in the forecasts of world oil price. Some models include assumptions that OPEC suppliers maximize a discounted stream of revenues net of costs (monopoly or cartel pricing). Others, such as the Department of Energy/Oil Market Simulation (DOE/OMS) model, include an implicit assumption that OPEC quantity trajectories are independent of oil prices or consumer response. When the two classes of models are exercised to project the impacts on world oil prices of reductions in U.S. oil imports, different answers are produced by the two: the DOE/OMS model typically projects a larger price impact than do the monopoly models. If the two classes of models were used to estimate the economic value to the U.S. of reducing oil imports, then two different answers would emerge.

The differences in price impacts seem to be the result of these behavioral assumptions. Monopoly pricing requires a production rate such that the demand elasticity facing the monopolist exceeds unity. In contrast, many analyses, including many conducted with the DOE/OMS model, embody an assumption that the demand elasticities facing OPEC are significantly smaller than unity. The price impact of a given quantity change is inversely related to the elasticity of demand for oil. Therefore, the structure of monopoly implies that these projected oil price impacts of reduced imports will tend to be smaller than the projected impacts based upon models such as DOE's Oil Market Simulation (8).

The above example does not tell us which results are more nearly correct, since the extent to which OPEC currently is behaving as a monopolist is subject to intense debate. However, through understanding the implicit contention point the debate over the benefits of reducing imports can be clarified.

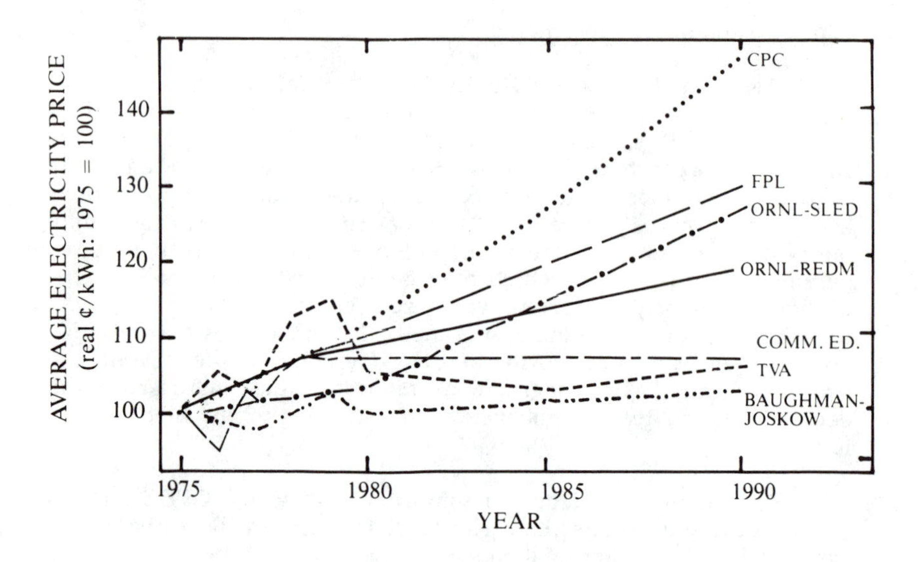

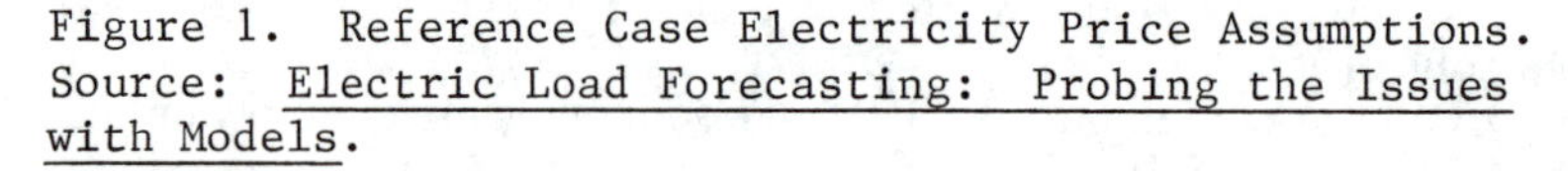
Figure 1. Reference Case Electricity Price Assumptions. Source: Electric Load Forecasting: Probing the Issues with Models.

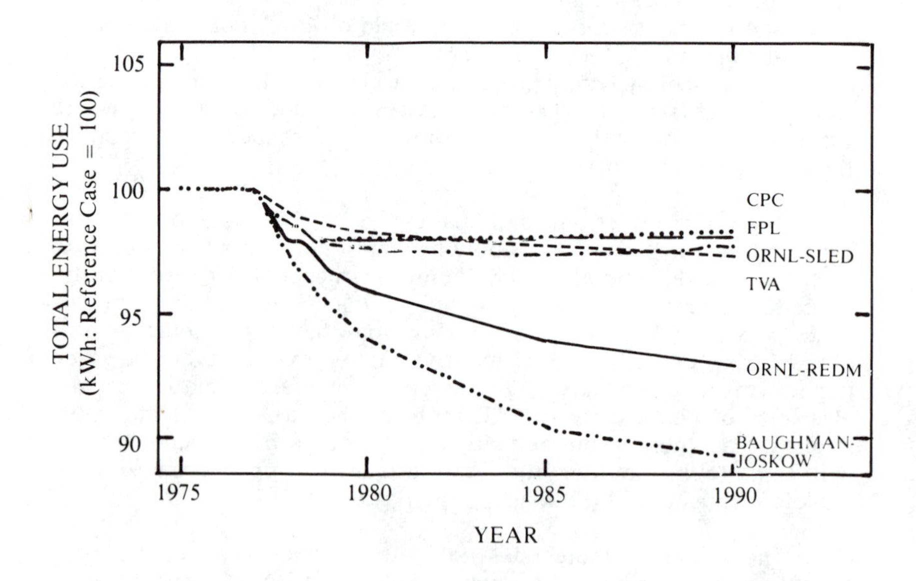

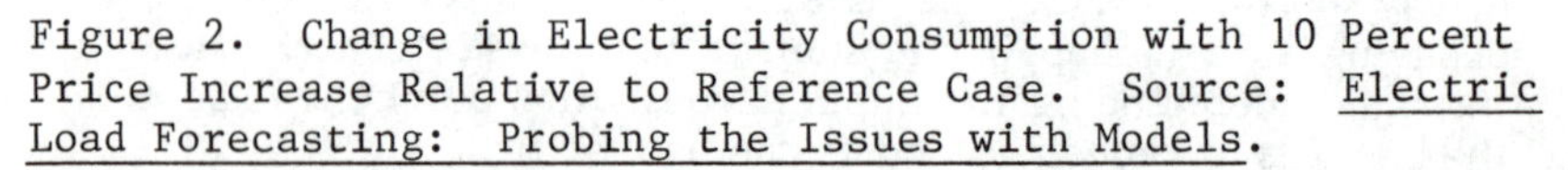
Figure 2. Change in Electricity Consumption with 10 Percent Price Increase Relative to Reference Case. Source: Electric Load Forecasting: Probing the Issues with Models.

Discussion of contention points may hasten diffusion of new ideas into many models, since this discussion may enhance basic understanding of the modeling issues. This flow of information into the body of analytical tools may be in the collective interest. However, the originator of a novel approach may be ambivalent about an instant diffusion of ideas into the competitor's models, especially if the only acknowledgment appears in a footnote of the mathematical addendum to Appendix 7.

A second example of model comparisons for clarifying contention points is EMF 3: "Electric Load Forecasting, Probing the Issues with Models" (3). In this study, the participants examined the responsiveness to electricity and natural gas prices of several electricity demand models. It was generally believed by the participants in the study group that electricity prices would continue to rise significantly (Figure 1). Thus, differences in the price elasticity of demand for electricity are fundamental to differences in forecasts of needed capital expansion for electric utilities. Figure 2 shows the results of model comparisons. One group of models had electricity price elasticities of demand equal to about -0.2, while two models--the Baughman-Joskow model and the Oak Ridge SLED model--showed elasticities of -0.7 and -1.0. The working group as a whole could not agree as to which behavior was closer to reality, although most working group members developed personal convictions. Vociferous disagreement arose on the correct resolution of this point. However, group members agreed on the value of actively debating and communicating those differences. Many hoped that publication of the study would facilitate and motivate a scientific debate, particularly within the electric utility industry, as to how significantly electricity price increases could be expected to reduce electricity demand growth.

The general category, clarification of disagreements, can be delineated into a number of specific areas as follows:

- Identification and communication of model limitations
- Identification of forecast uncertainty
- Enhancement of system understanding
- Guidance for future research

An examination of these specific benefits follows.

Identification and Communication of Model Limitation

Every mental and formal model has limitations. Limits of application should not be embarrassments. However, the failure to identify and communicate these limitations increases the probability that the model will be misused. Identification of the limitations of a model is the first necessary step for avoiding such misuse. This can be accomplished through discussion of equations, through examination of sensitivity tests, and so on. Even when the limits are apparent to the developer of a model, they may not be discernible to those who have not been closely associated with the specific modeling project.

For example, reduced form econometric models of gasoline demand were developed using data that predate the corporate average fuel efficiency standards on new cars. Thus, the structural change associated with the recent implementation of these standards implies that those reduced form models will tend to both over-estimate gasoline demand and over-estimate the role of higher gasoline prices in reducing demand (4, 9). What had not been a limitation to these models now has become a subtle, but important and poorly understood limitation. Comparison of the current generation of structural gasoline demand models with the reduced form econometric models can crystallize understanding of this type of limitation.

Model benefits depend upon communicating the limitations to other modelers and model users. The essence of truth-in-modeling is disclosure: disclosure of methods, assumptions, data, approximations, sensitivity results, insights, advancements, and embarrassments. The converse "untruth in modeling" is normally associated with a failure to allocate sufficient resources to the examination of models and communication of those ideas. Full disclosure is a most difficult task faced by modelers. Lack of resources is probably the single most critical barrier to full communication. In addition, political issues, competition, gamesmanship, and contractual obligations encourage the modeler to bypass the discussion of weak points and to concentrate on strengths and insights.

Serious effort on the part of modelers and model users is needed to overcome the barriers. For example, model assessment by third parties has certain advantages; see for example (11). Third-party assessment often provides an opportunity for the modeler to step back and to ask broad questions about key limitations. The assessor is often not initially aware of the complete structure of the model and thus has the opportunity to provide a new perspective. Since professional recognition for assessment activities is usually associated with the identification and communication of the limits of a model, third-party assessment activities may be quite effective.

Group exercises concentrating on the comparison of models' aggregate behavior are also useful. Such exercises can identify and solidify understanding of model capabilities and limitations and can facilitate communications to a broad audience of potential model users. Since all models have limitations and since both strengths and limitations of models being compared can be discussed, the modelers collectively benefit from the truth-in-modeling encouraged by such an exercise.

Identification of Forecast Uncertainty

Model comparisons make us more humble about our ability to forecast anything, using either mental or formal models, with precision. More positively, model comparisons reveal the full range of uncertainty of any forecasts. In modeling, three types of uncertainty exist: in inputs, in parameters, and in structure. These sources of uncertainty can each be addressed through model comparisons.

In order to examine the uncertainties of a forecast, modelers will often vary a set of critical input assumptions to the models over a range of values and observe changes in outputs. For example, this is a procedure used in the 1977 Annual Report to Congress (15) by the Energy Information Administration. Such sensitivity tests can provide a mapping from uncertainties in critical inputs to uncertainties in the output projections. These sensitivity tests allow researchers to identify the pivotal uncertain data elements to be input to the model. A high degree of uncertainty about a pivotal data element will severely limit the precision of any forecast and should motivate an effort to better estimate this element. Thorough scrutiny of these uncertainties through model comparison can provide more precise estimates of uncertainty.

A second element of uncertainty that is less frequently examined is the relationship between the parametric estimates and the real system being modeled. Typically, parameters are estimated using either statistical or engineering data methods along with professional judgment. Uncertainty in outputs associated with uncertainty of these parameters can be examined using a Monte Carlo simulation, randomly choosing alternative possible values of those parameters. Such simulations conceptually allow a mapping from uncertainties in critical parameters to uncertainties in the outputs.

Both types of uncertainties could be examined simultaneously, through use of Monte Carlo simulation techniques. As long as parameter uncertainties and input uncertainties are not correlated, the joint output uncertainty will be greater than the measurement obtained by allowing only one of these to vary. Therefore, examination of only input uncertainty will lead to underestimates of the real uncertainty about outputs.

The third source of uncertainty can be thought of as structural uncertainty. Inevitably forecast errors exist because the modeling process always results in a model structure that differs from reality, often in important ways. And often there is great uncertainty as to which of several structures best approximate reality. For example, there may be alternative paradigms utilized to forecast the impacts of policy choices. The debate in economics between the Macroeconomists influenced by Keynesian concepts and the Monetarists provides such an example. Even more dramatic are the paradigm differences among the so-called radical economists and the neoclassical economists. At a less basic level are the differences cited earlier among the alternative conceptual structures underlying models of world oil pricing. These differences imply a deep uncertainty that goes beyond differences in inputs or parameters.

Forecast uncertainty stemming from structural uncertainty is poorly understood and difficult to analyze. It is even unclear how one could conceptually define a measure of structural uncertainty since the set of feasible alternative structures cannot be observed and can only be vaguely imagined.

Some understanding of the degree of structural uncertainty can be obtained by comparison of several models that incorporate different representations of the same system. At least such comparisons help one to develop a more profound humility about the quality of our knowledge of the world than would be developed if forecasts from only one model were used.

Model comparisons do not fully answer the questions posed by structural uncertainty. However, they do provide a starting point for future analysis of uncertainty, since they identify the areas of uncertainty, explore the quantitative significance, and motivate further analysis.

Enhancement of System Understanding

Another important benefit resulting from a formal comparison is increased understanding of the system being modeled. There are a number of examples in the energy area. Models by Edward Hudson and Dale Jorgenson (10) and Ernie Berndt and David Wood (17) have forced a reevaluation of the relationships between energy prices and the rate of capital formation in the U.S. economy. The debate between Berndt/Wood and James Griffin (14, 18) concerning their models of capital formation has led to an enhanced understanding of capital-energy interrelationships.

Work produced from the first Energy Modeling Forum working group (1), under the leadership of Bill Hogan, reexamined the popular belief that reductions in energy would inherently lead to almost proportionate reductions in economic growth. Attempts to compare and explain differences between economic impacts projected using various energy-economy models provided insight for Bill Hogan's and Alan Manne's paper, "Energy-Economy Interactions: the Fable of the Elephant and the Rabbit?"(6) This paper provided important insights about the relationship between the elasticity of demand for energy and the economic impact of reduced energy availability.

Guidance for Model Selection

The last category of model comparison benefits is guidance in selection of appropriate models for specific applications. For most analysis, it is important to choose a model that matches the desired function because the choice of model may greatly influence the results obtained.

Analysis of the impacts of an oil import quota demonstrates this idea. The PILOT model (19), which uses a very small elasticity of demand, would show a very large impact of a quota on GNP, while the Hudson-Jorgenson model (10), which has a much higher demand elasticity, would show a much smaller impact. The projected impacts on world prices obtained by utilizing a formal or mental model, such as the Salant model (16), which envisions OPEC as a profit-maximizing monopoly or cartel, will be very different from those obtained by a model such as the DOE/OMS, which implicitly views the world oil market as competitive. The latter class of model would imply that the quota, by reducing the demand for world oil, would reduce oii prices. The former class--monopoly models--might indicate that a broadly applied quota which reduces the elasticity of demand facing OPEC will motivate price <u>increases</u>.

The extent to which the model comparison exercise should be carried is a function of the importance of the decision or policy and the potential influence of analyses on the ultimate decision. Clearly, when the issue is as important as the imposition of oil import quotas by the U.S. and its allies, it is inexcusable to fail to use a number of formal models, comparing their results with those obtained using purely mental models. Failure to do so may result in decisions based on an inadequate information base.

For some decisions, however, it is not economical or feasible to conduct an adequate model comparison exercise. The urgency of a decision may require expedience or the sensitivity of an issue may demand a minimal number of people involved. In these cases it is valuable to have an information base about the relative behavior of several models. This information base can allow a shortcut to understanding the assumptions that are implicitly made by choice of a specific model. The informed user can adapt the results from those models that most closely correspond to his or her interpretation of the evidence. If the model has already been selected (which is often the case with "inhouse" model development), the model user can adjust the results according to his or her judgment and knowledge of other models.

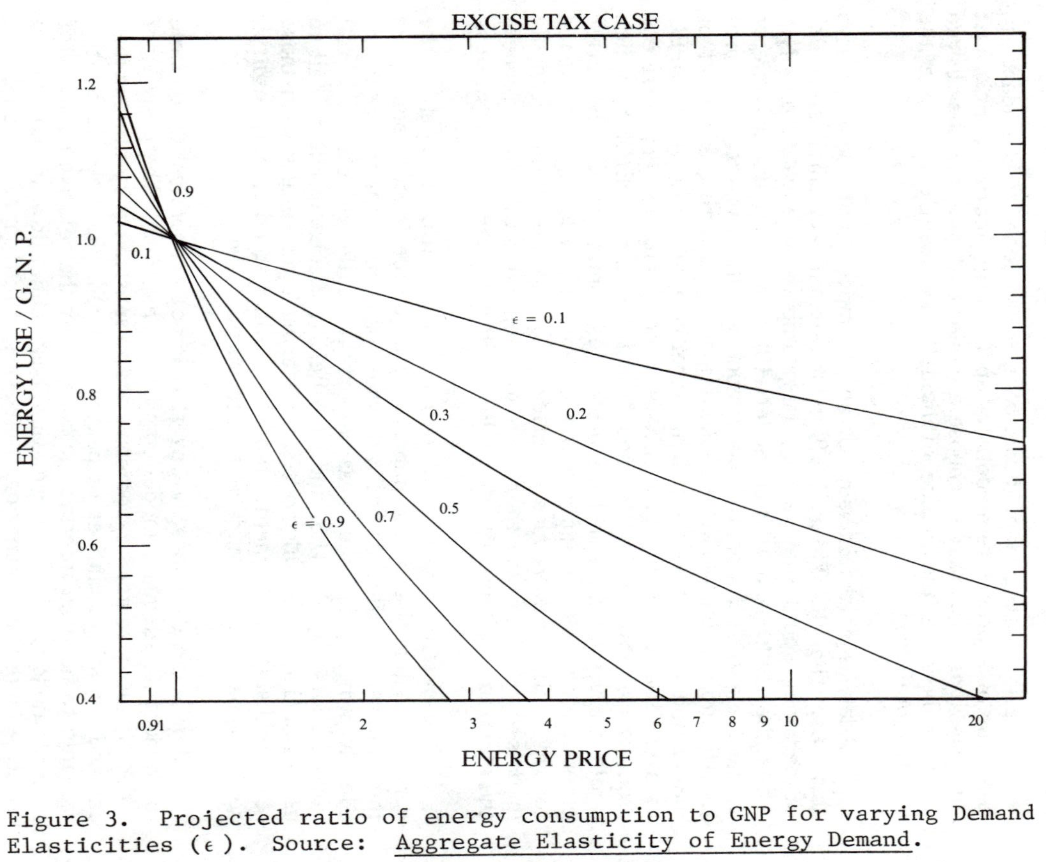

Figure 3. Projected ratio of energy consumption to GNP for varying Demand Elasticities (ϵ). Source: <u>Aggregate Elasticity of Energy Demand</u>.

For example, if it were necessary to project the use of energy to the year 2000, model selection could be influenced by knowledge of the responsiveness of different models to higher energy prices. The degree of responsiveness is quantified by the aggregate elasticity of demand for energy implicit in the model. This elasticity is commonly measured at one of three points in the supply chain: primary energy (at the wellhead or minemouth), secondary energy (at the refinery gate or electric utility busbar), or delivered energy (at the gas pump or meter). Figure 3 shows how the projected ratio of energy consumption to GNP varies with aggregate energy price for various demand elasticities measured at the secondary level. For a given future price of energy, the projected energy/GNP ratio depends upon the aggregate elasticity of energy demand implied in the model. The estimate of aggregate elasticity of energy demand will fundamentally shape projections. The importance of this measurement of elasticity for forecasting energy consumption encouraged the Energy Modeling Forum to compare the elasticities of sixteen energy demand models used in their EMF 4 study (4).

In Table 1, the models have been classified on the basis of the parameter estimation procedure used--statistical, engineering, or judgmental. Aggregate elasticities of demand at the secondary level are shown for the separate consuming sectors of the various models. Such a tabulation can be useful for guiding users in the selection of a model for specific analyses. Presumably, that choice will be based upon the evidence, as interpreted by the user, of the responsiveness of energy demand to prices. On a more cynical note, such a table could be used to select that model which gives the answer most convenient for the user. Even for such cynical purposes, the existence of the data-base gives both the primary user of the model and other users--adversaries, allies, and impartial observers--a firm understanding in order to debate specific forecasts, analyses, and conclusions.

Methods of Model Comparison

Implicit in the preceding discussion of model comparison benefits are some notions as to how those comparisons might be conducted. This section makes explicit those implicit notions. Four broad categories of model comparison are discussed: methods and equations, forecast, aggregate behavior, and model regeneration. In addition, "modeling-the-model" helps in structuring and communicating these comparisons.

Methods and Equations Comparisons

The most frequently developed model comparisons typically describe the methodologies incorporated in the various models and compare and contrast several equations from the models. Such

Table 1

Aggregate Elasticity of Energy Demand

SECTOR	STATISTICAL	ENGINEERING	JUDGMENTAL
RESIDENTIAL	GRIFFIN OECD .9 MEFS .5 PINDYCK 1.0	BECOM .6 HIRST RESIDENTIAL .4	
RESIDENTIAL/ COMMERCIAL	BAUGHMAN-JOSKOW .8 BESOM/H-J .7 MEFS .5	BECOM .5	EPM .5
COMMERCIAL	MEFS .5	BECOM .3 JACKSON COMMERCIAL .4	JACKSON .4
COMMERCIAL/ INDUSTRIAL	GRIFFIN OECD .3 PINDYCK .7		
INDUSTRIAL	BAUGHMAN-JOSKOW .4 BESOM/H-J .5 MEFS .2	ISTUM .2	EPM .7
TRANSPORTATION *	FEA-FAUCETT .1 GRIFFIN OECD .5 BESOM/H-J .2 SWEENEY AUTO .5 WHARTON MOVE .2 PINDYCK .5 MEFS .3		EPM .4
ALL SECTORS	BAUGHMAN-JOSKOW ** .6 GRIFFIN OECD .5 MEFS .3 PINDYCK .7 BESOM/H-J .4		ETA-MACRO .2 FOSSIL 1 .1 FOSSIL 1/ CONSERV. .2 PARIKH WEM .1 EPM LL) .6

* The FEA-Faucett, Wharton, and Sweeney results are for automobile gasoline only. These are 15 year elasticities.

** Excludes the transportation sector

comparisons are often based only upon the existing model documentation, possibly enhanced by discussions with the model developers. Often, they can be recognized by a title: "A State-of-the-Art Review of . . . ." Model comparisons of this type are also familiar to readers of academic literature, since a summary of previous work is a normal component of academic journal articles. In the MIT Energy Laboratory this type of model assessment is incorporated in the "Overview Assessment" phase of analysis (11). This type of comparison is the "descriptive" phase where the predominant methodology is common sense and professional judgment.

There are three basic elements of Methods and Equations comparison. The first is a discussion of the basic structure and underlying theory of the models. The relationship between the theory and its implementation into the model's structure is a crucial element of this component. The goal is to pinpoint important differences between theories and between various interpretations of a theory.

The second component evaluates the techniques of measurement and estimation used in the models. This allows the analyst to ponder possible biases in model behavior. More importantly, it allows judgment of the dependability of parameter estimation.

A key aspect of methods and equations comparison that improves communication is the reproduction of key equations and parameter values. For models having thousands of equations, this step may be impossible for more than a few of those equations. However, even in the most complicated models, often only a limited set of equations need to be examined carefully. Most of the equations are necessary only for details.

A part of a methods and equations comparison is a delineation between the new model and the "state-of-the-art" knowledge. Such comparisons provide a basic perspective for the subsequent comparisons. For example, at the MIT Energy Laboratory a review of the literature occurs early in the assessment procedure. This helps to place a specific model within a context of alternative models or approaches to an issue.

Forecast Comparisons

The second category of model comparison involves contrasting the predictions of the models. Forecast comparisons can be conducted on four levels. The first level is essentially a descriptive comparison of the array of forecasts. Since it is a cursory analysis, it alone seldom gives insight into the source of differences.

The second step attempts to adjust the various forecasts to a common basis. Forecasters generally will have used widely differing assumptions about the future values of the driving variables. Knowledge of those differing input values and knowledge of the mapping from input changes into output changes together allow adjustments of the forecasts to common bases.

The third and more complete level requires operation of at least one model. It then is possible to choose input values identical to those of a published forecast so as to allow a baseline comparison. This procedure was used, for example, in the early development of the Project Independence Evaluation System (PIES) at the Federal Energy Administration. Projections were made for prices remaining at pre-embargo levels, and these were compared against the Dupree-West forecasts published by the U.S. Department of the Interior (20) before the steep rise in energy prices. It was reassuring that the PIES projections fairly closely matched the Dupree-West forecasts for a low price situation. More importantly, errors in the model were discovered early through these comparisons.

The fourth phase of forecast comparison involves examination of all the models through operation with standard data. This assures that forecast variations stem only from model differences and not from the specific data inputs. The Energy Modeling Forum typically uses this procedure.

Another type of comparison that is closely related to the forecast procedure is called a "backcasting" comparison. The procedure requires the operation of each model with historical standardized data. Then the forecast "predicts" an event that has already occurred.

The Conference on Econometrics and Mathematical Economics (CEME), sponsored by the National Bureau of Economic Research and the National Science Foundation (21), provided a forum for the comparison of macroeconomic models. Inputs have been set at historical values in order to backcast the macroeconomic behavior that actually occurred. Summary statistics have been compared, describing how well the various models performed in these structured tests. Such backcasts, when feasible, provide valuable information.

Forecast comparisons, by themselves, do not necessarily expose the precise reasons for differences in the predictions. The information derived from the procedure may be of limited value when considered by itself. As a component of model comparison, the forecast exercise can prove to be crucial.

Aggregate Behavior Comparison

Most models contain many complex relationships and provide detailed projections, often disaggregated by sector, product, location, and so on. These detailed relationships are often important for precise delineation among policy options or for projections of differential impacts among locations, sectors, or products. But these detailed relationships may not be important for understanding the more aggregate outputs from the models, in fact, may help hide the more basic forces. Therefore, it is often valuable to develop comparisons among the broad aggregate input-output relations of the various models. This is the process of aggregate behavior comparison.

For example, some oil supply models project production of crude oil by region, by specific gravity of the oil, and by sulfur content. Inputs required are current and future world oil prices, tax structure, detailed price control rules, and so on. Aggregate behavior comparisons may ignore this detail and only examine the total lower-48-state oil production rate and the influence of a proportionate increase in all oil prices on this total.

Several examples of aggregate behavior comparisons can be cited. The CEME has developed aggregate behavior comparisons of macroeconomic models, focusing most attention on the various multipliers--government expenditure, tax, investment. The MIT Model Assessment Laboratory also incorporates this type of exercise in their "In-depth" phase of model assessment. The Energy Modeling Forum studies include extensive efforts to characterize the aggregate behavior of models, based upon common scenarios run by each modeler. For instance, in EMF 4: Aggregate Elasticity of Energy Demand (4), energy prices and economic growth assumptions were standardized for each of nine scenarios, predicting different rates of change of energy prices. The aggregate elasticity of energy demand was characterized for each model, based upon the outputs. Results appeared in Table 1. In EMF 5: U.S. Oil and Gas Supply (5), the responsiveness of total oil supply and total gas supply to assumptions about geological potential, oil price trajectories, or policy changes was characterized. Other examples include characterization of the influence of appliance efficiency standards, of average price and price structure, or of natural gas prices on electricity demand (EMF 3; see Figure 2); or the reductions in economic growth associated with restrictions on the availability of energy (EMF 1) (1).

Aggregate behavior comparisons normally include three key elements:

- Standardization of inputs for the models
- A set of structural sensitivity tests

- Summary characterization and quantification of key aggregate relationships

In contrast to the methods and equation comparison, the aggregate behavior of the models can be compared, even though the models may have different structures. For example, we can compare the response of oil supply to price in an intertemporal optimization, linear programming model (LORENDAS), a mixed econometric-engineering model (AGA-TERA), and a purely econometric model (Rice-Smith), even though these models have vastly different structures (5). However, it seems impossible to compare these models on an equation-by-equation basis.

Model Regeneration

The final mode of comparison may be called model regeneration. Critical relationships are examined by systematically replacing one component of the model (parameter set, equation, or system of equations) with an alternative version of the same component. By this procedure a family of models is generated based on one primary model. Changes in the model behavior then can be examined. An example of model regeneration is the MIT Energy Laboratory "Counter-Analysis" activities used in the in-depth evaluation of the Baughman/Joskow Regionalized Electricity Model (11). While the most costly mode of comparison, model regeneration exercises offer perhaps the greatest potential for model assessment of any method discussed above.

Modeling-the-Model

The final exercise of model comparison, "modeling-the-model," provides a means to consolidate the results from the other comparisons and a forum for interpretation of the results. The modeling-the-model idea requires the development of a simple model characterized by a very limited number of key parameters. This method can roughly mimic the actual model in order to illustrate, analyze, and communicate the structure of complex models.

For example, for EMF 1, Bill Hogan and Alan Manne postulated a simple constant-elasticity-of-substitution production function between energy and all other inputs to the economy. Two parameters were free, the value share of energy in the economy and the elasticity of substitution between energy and other productive factors. Differences in the aggregate elasticity of substitution implicit in the model led to fundamental differences in the projected economic growth reductions associated with restrictions on energy availability. It was shown that the aggregate elasticity of substitution could be closely approximated by the aggregate elasticity of energy demand discussed in previous sections of this paper. The relationship between energy availability and economic growth for various aggregate demand elasticities (ranging from 0.1 to 0.9) is illustrated in Figure 4. While this very simple structure could not capture many elements of the more complicated models, it captured perhaps the most important three issues. This simple model-of-models allows individuals to understand the role of elasticity of substitution in governing model-based projections of energy-economy interactions.

The modeling-the-model technique has been used within EMF 5 to emphasize the importance of the shape of the finding curve, the mathematical relationship between new discoveries of oil per additional unit of exploration and cumulative exploration for oil. The assumed shape of the finding curve can greatly influence the model's projections of the ultimate quantities of oil discovered and the price elasticity of the cumulative discoveries.

Within many models is a maintained assumption that the discoveries per unit of additional exploration (feet drilled, wells completed, etc.) is an exponentially declining function of cumulative exploration. This assumption of an exponential finding curve implies that the discoveries per additional well are always proportional to undiscovered reserves. A simple alternative "model-of-the-models" embodies an assumption that the discoveries per additional exploration are always proportional to undiscovered reserves, raised to the power B. If B=1, then the finding curve is exponential, corresponding to most of the complex models. Low values of B imply that the productivity of drilling is almost independent of remaining reserves, while a large value of B implies that the productivity of drilling declines sharply as additional reserves are found. For all values of B the postulated finding curve can be calibrated to the same estimates of initially undiscovered resources and initial rate of oil discoveries per well.

Figure 5 shows results from the simple model, plotting the fraction of resources which will be cumulatively discovered under a profit-maximizing regime as a function of the "relative price." This relative price is the value of a discovery relative to the initial exploration cost per discovery. Relative price is roughly proportional to the price of oil. The initial exploration cost per discovery is the cost per well divided by the the initial discoveries per well. As B ranges from 0.6 to 3.0, the fraction of resources cumulatively discovered by a profit maximizing competitive industry drops sharply, varying by a factor of 2.0 or more for low relative price. Figure 6 plots the elasticity of cumulative discoveries with respect to the value of a discovery for various B's. This supply elasticity drops as B decreases, varying by a factor of 3.0 or more as B varies from 0.6 to 3.0.

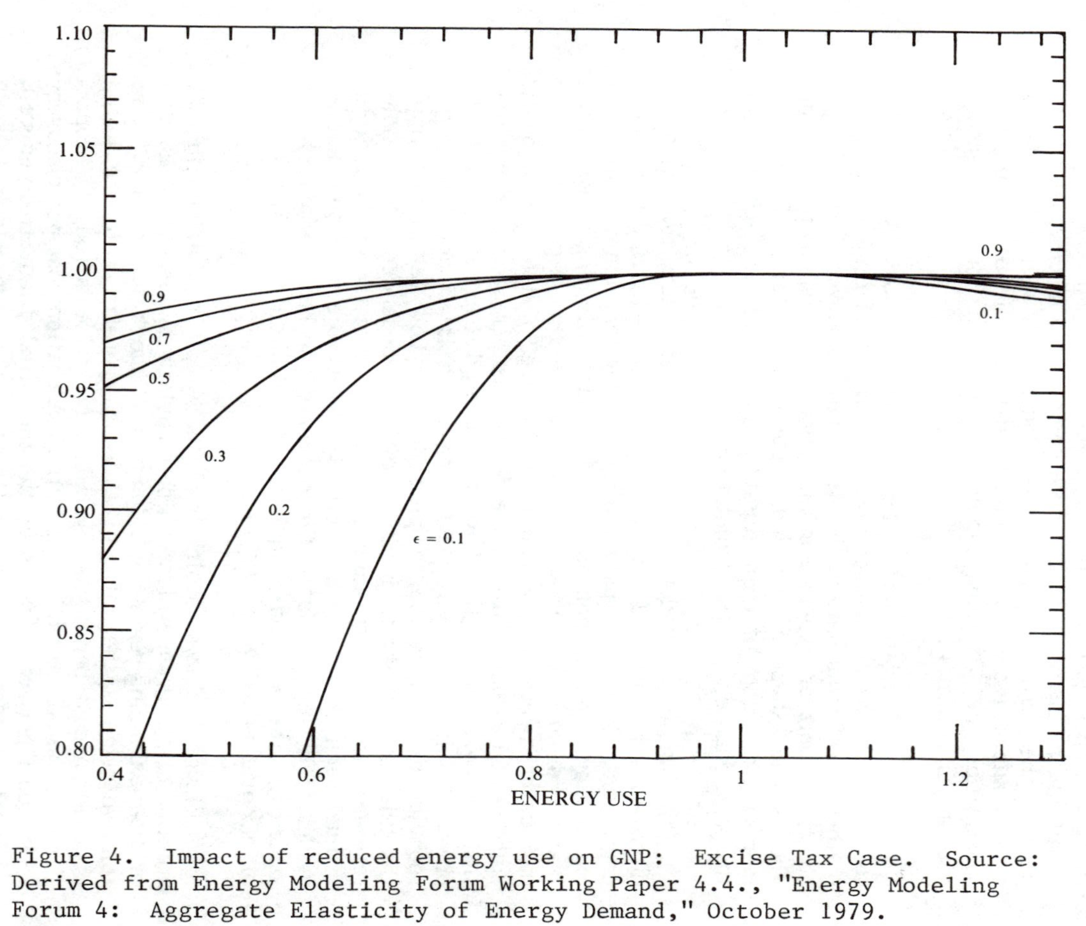

Figure 4. Impact of reduced energy use on GNP: Excise Tax Case. Source: Derived from Energy Modeling Forum Working Paper 4.4., "Energy Modeling Forum 4: Aggregate Elasticity of Energy Demand," October 1979.

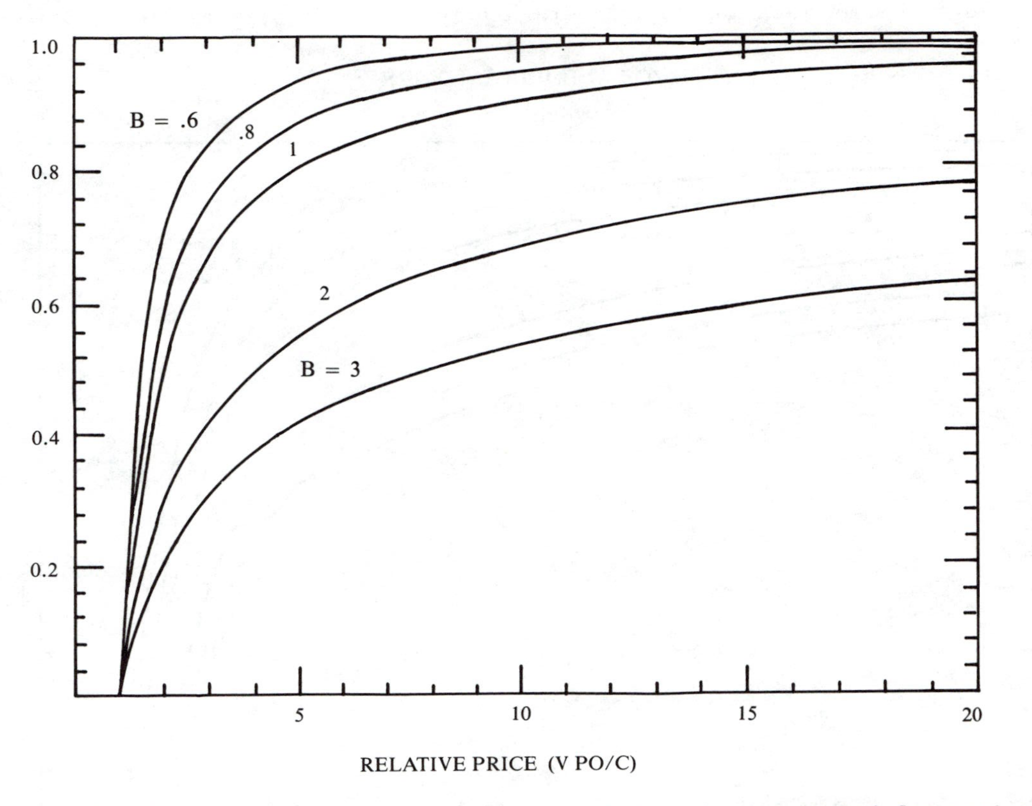

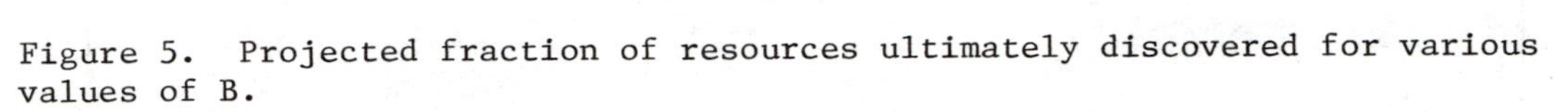

Figure 5. Projected fraction of resources ultimately discovered for various values of B.

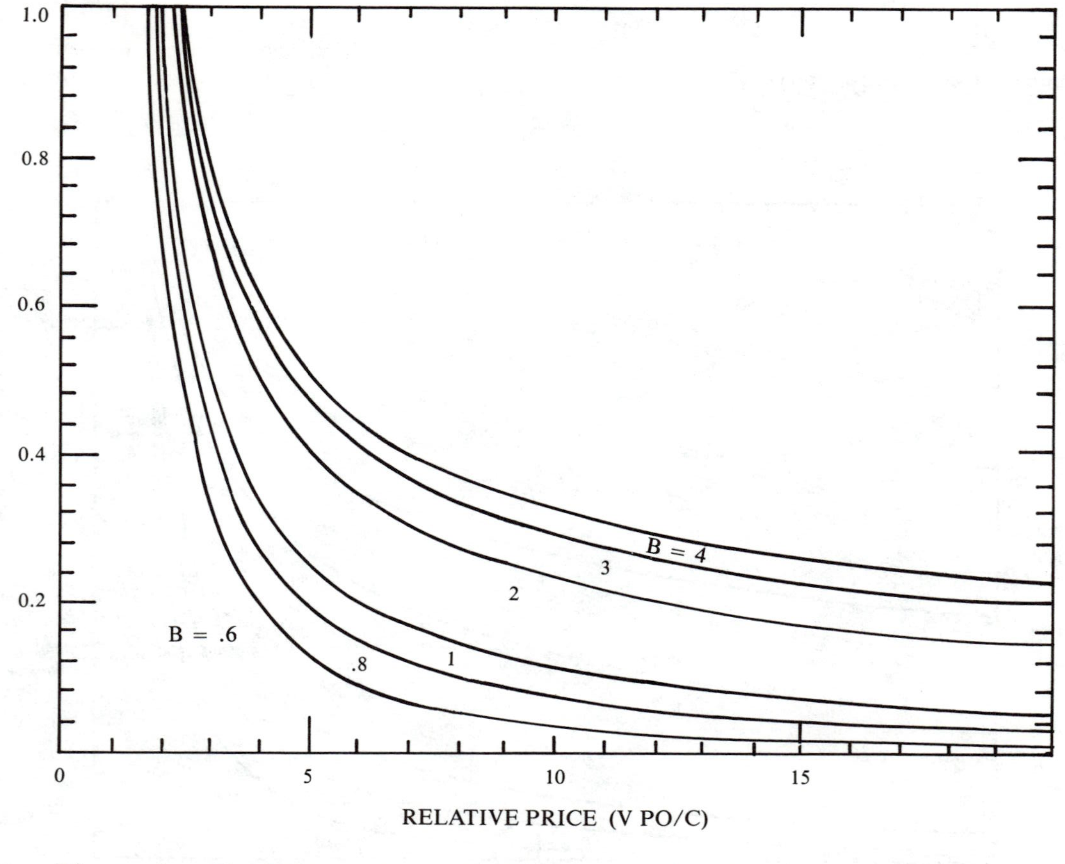

Figure 6. Elasticity of cumulative discoveries for various values of B.

Both cumulative supply and the price elasticity of supply will depend critically upon B. How well is the value of B known? Current information does not allow one to limit B between 0.6 and 3.0. Current research results certainly do not strongly justify a choice of B equal to unity. Thus, in this case, the simple model (1) allows vivid interpretation of implicit assumptions embedded within the models; (2) highlights the finding curve shape as a critical determinant of long-run oil and gas supply; and (3) suggests the potential value of research into finding curve shapes.

Conclusions

Formal model comparison activities can provide important benefits, ranging from simple identification of coding errors to greater fundamental understanding of the system being examined. The benefits obtained may vary systematically with the comparison methods used.

Methods and equations comparisons can assist in understanding the underlying theories employed and can enhance understanding of the system being studied. Yet this approach, taken alone, can be deceptive by overemphasizing methodological differences that may be insignificant quantitatively or by underemphasizing numerical or structural differences that may lead to fundamental differences in aggregate behavior.

The second and third modes of analysis, forecast and aggregate behavior comparison, are particularly useful in developing understanding of which issues are quantitatively significant. However, when taken alone, these methods can produce a cursory analysis. Numbers without analysis of the source of these numbers may provide little insight into the dependability of a model or the priorities for its improvement.

Model regeneration exercises can provide the most detailed understanding of the relationships between the parts of a model and its whole and can provide a means of testing the implications of alternative theories. However, regeneration exercises, taken alone, can waste vast amounts of computer and human time without producing basic understanding.

Modeling-the-model can focus discussion on the few most fundamental issues and can provide a method of disseminating results. However, unless the simple results are related to the aggregate behavior of the models, this approach may be little more than a mathematical exercise.

In short, the various methods each have severe limitations when taken alone. When taken together they can be highly complementary. Table 2 presents rough summary judgments about

Table 2

Model Comparison Methods and Their Benefits

	METHODS AND EQUATION COMP.	FORECAST COMP.	AGGREGATE BEHAVIOR COMP.	MODEL GENERATION	"MODEL-THE-MODEL" COMPARISON
Identification of Errors	G	G	G	E	P
Clarification of Disagreements	G	P	G	G	E
Identification & Commun. of Limits	G	G	E	G	P
Identification of Uncertainty: Data Inputs	P	G	G	G	G
Parameter Estimates	G	G	E	E	E
Structural	P	P	G	G	P
Guidance for Research	G	P	E	G	G
Enhancement of System Understanding	E	P	E	G	E
Guidance in Model Selection	G	G	E	G	G

P = Poor G = Good E = Excellent

the relations between comparison methods and benefits. This tentative summary is open to severe debate, criticism, and revision. The basic point is that one can expect a systematic relationship between the modes employed and the benefits obtained. For example, modeling-the-model may be useless for the identification of coding errors while model regeneration could be invaluable. Conversely, clarification and communication of disagreements can be organized through modeling-the-model whereas forecast comparisons alone will give little insight.

Three problems identified in the Introduction provided much of the motivation for this paper: (1) communication among model developers is surprisingly limited; (2) models remain as black boxes, especially to potential users; and (3) models are seldom used in a comparative mode to address policy or planning questions. These problems are a result of lack of understanding and communication. Comparative model studies can alleviate some of the difficulties associated with these problems. The current trend toward increased model comparison could lead to more effective model application of models to policy issues. This type of endeavor is worthy of continued encouragement.

Acknowledgments

I would like to gratefully acknowledge, without implicating, helpful comments and critiques from Gary Ardell, Patrick Coene, Douglas Finlay, David Froker, Douglas Logan, and John Weyant, and editorial assistance from Dorothy Sheffield and Pamela Rosas. I would especially like to thank Tricia Rigot for her extensive research and writing assistance.

References

1. Energy Modeling Forum, Energy and the Economy, EMF 1, Stanford University, Stanford: September 1977.

2. Energy Modeling Forum, Coal in Transition: 1980-2000, EMF 2, Stanford University, Stanford: July 1978.

3. Energy Modeling Forum, Electric Load Forecasting: Probing the Issues with Models, EMF 3, Stanford University, Stanford: April 1979.

4. Energy Modeling Forum, Aggregate Elasticity of Energy Demand, EMF 4, Stanford University, Stanford: August 1980.

5. Energy Modeling Forum, "U.S. Oil and Gas Supply Summary Report," Working Paper, EMF 5, Stanford University, Stanford: November 1979.

6. Hogan, W.W. and A.S. Manne, "Energy Economy Interactions: the Fable of the Elephant and the Rabbit?," in Hitch, C.J. (ed.), Modeling Energy-Economy Interactions: Five Approaches, Resources for the Future, Washington, D.C.: 1977.

7. Greenberger, M., M.A. Crenson, and B.L. Crissey. 1976. Models in the Policy Process: Public Decision Making in the Computer Era. New York, Russell Sage Foundation.

8. William Hogan. Verbal Communication.

9. Sweeney, J.W., "Energy Policies and Automobile Use of Gasoline," in Hans Landsberg, ed., Selected Studies on Energy. Background Papers for Energy: The Next Twenty Years, Ballanger Publishing Company, Cambridge, Mass., 1980.

10. Hudson, E.A. and D.W. Jorgenson, "U.S. Energy Policy and Economic Growth, 1975-2000," Bell Journal of Economics and Management Science, Vol. 5, No. 2, Autumn 1974, pp. 461-514.

11. M.I.T. Model Assessment Group, "Independent Assessment of Energy Policy Models: Two Case Studies," Report no. 78-011, Cambridge, Massachusetts: M.I.T. Energy Laboratory and Center for Computational Research in Economics and Management Science, 1978.

12. Sweeney, J.L. and J.P. Weyant, "The Energy Modeling Forum: Past, Present, and Future," Energy Policy, 1979. Also, Hogan, W.W., "The Energy Modeling Forum: A Communication Bridge," Paper presented at the Eighth Triennial IFORS Conference, Toronto, June 20, 1978. Also, Hogan, W.W., "Energy Models: Building Understanding for Better Use," Paper presented at the Second Lawrence Symposium on Systems and Decision Sciences, Berkeley, October 3-4, 1978.

13. Crissey, B.L, "A Rational Framework for the Use of Computer Simulation Models in a Policy Context," Ph.D. dissertation, Baltimore: The Johns Hopkins University, November 1975.

14. Berndt, E.R. and D.O. Wood, "Engineering and Economic Interpretation of Energy-Capital Complementarity," American Economic Review, June 1979, Vol. 69, No. 3, pp. 242-254.

15. Energy Information Administration. Annual Report to Congress, 1977.

16. Salant, S., "Imperfect Competition in the World Oil Market," Lexington Books, October 1981.

17. Berndt, E.R. and D.O. Wood, "Technology, Prices, and the Derived Demand for Energy," Review of Economics and Statistics, Vol. 58, No. 1, (February 1976) pp. 1-10.

18. Griffin, J.M. and P.R. Gregory, "An Intercountry translog Model of Energy Substitution Responses," American Economic Review, December 1976, Vol. 66, No. 5, pp 845-857.

19. Dantzig, G.B., T.J. Connolly, and S.C. Parikh, "Stanford Pilot Energy/Economic Model," Electric Power Research Institute, EPRI EA-626, May 1978.

20. Dupree, W.E., and J.A. West, "United States Energy through the Year 2000," U.S. Department of the Interior, December 1972.

21. Klein, L.R. and E. Burmeister, Econometric Model Performance, University of Pennsylvania Press, 1976.

Charles C. Holt

13. Quantitative Modeling: Needs and Shortfalls for Energy Analysis

Two recent news events dramatically symbolize the theme of this paper. First, the war between Iraq and Iran threatens much of the world's oil supply and precipitates the urgent need for public and private, immediate and long-term actions. Second, award of the Nobel prize in economics to Lawrence Klein honors a major contributor to modeling relevant to forecasting and policy analysis. Unfortunately the present state of the art lacks the practical tools that are so urgently needed. This paper analyzes the needs for quantitative analysis and our present capabilities to build and apply models in decision-making, particularly in the context of energy problems (1).

Energy Problems and the Need for Models

Since abundant, easily accessible oil deposits were discovered in 1859 in Pennsylvania and later in Texas, Louisiana, and the Middle East, the United States in particular and industrial countries in general have fallen heir to the benefits of a long period of cheap oil and gas energy. More than any other country the United States responded to this period of cheap energy by building a high standard of living based on an energy-intensive technology that literally was built into the designs of its manufacturing machinery, electric power grid, transportation system, buildings, and household appliances.

The costs of finding and extracting these abundant but exhaustible natural resources were so low that their prices tended to be unstable in the face of demand fluctuations. This triggered U.S. governmental intervention at both the federal and state levels to prevent price excursions that were thought to be excessively high when business was booming or excessively low when business was depressed.

Now that the most accessible petroleum resources in the U.S. have been exhausted, putting production into decline, and the oil exporting countries have organized their collective bargaining power into the OPEC cartel to more than quadruple oil prices, the U.S. faces a drastic adjustment of its production and consumption patterns, including the lowering of its standard of living through the worsening of its terms of international trade with OPEC. The power of the cartel will ultimately be limited when the oil price becomes so high that it induces an increase in the production of domestic oil and the substitution of alternate energy sources. Unfortunately the required adjustments are so large, costly, and time consuming, the environment so uncertain, and the indirect effects so powerful that we face an extended period of inflation and/or unemployment and military insecurity arising from threats to energy supplies.

The costs, risks, and externalities are so large that governments cannot sit idly by and wait for the market to adjust. Continued increases in the real price of energy depend on oil exhaustion, which is fairly predictable, and the unpredictable interactions between the cartel power of the exporting countries and the counter-actions by the importing countries. The required adjustments and especially their timing are both critical and uncertain. In addition, political uncertainties surround atomic energy, and technological uncertainties and heavy capital requirements cloud the development of alternate energy sources. Thus if timely action is to be taken, government may need to underwrite some of the risk which otherwise could inhibit private investments.

Most energy production and consumption in the U.S. occurs in the private economy, which is motivated by self interest and largely regulated by markets. However, high energy costs can seriously affect household budgets for such necessities as house heating, electricity, and transportation to work. Clearly energy is involved not only in the issues of economic and technological efficiency, but also in the issues of equity and income distribution. Solving energy problems will thus require the intricate interplay of public policy with the private economy.

The broad outline of the required adjustments to higher energy prices can be foreseen. Energy-intensive technologies will be replaced by energy-saving technologies but large capital investments will be required first. Substantial expenditures must be made on research and development to produce the new product and process designs. Energy resources that previously were too expensive to compete with oil now will become competitive, resulting in the gradual emergence of a new industrial mix that will shift the location of energy sources, industrial plants, population, and transportation load and mode. Since many of the industry adjustments will be region specific, state and local governments will face substantial problems of accelerated growth or decline occurring in new patterns.

In this complex environment, governments and businesses need to coordinate their activities. Since social and private interests will need to be reconciled, public-private interactions will increase. Socially desirable activities in the private sector will be encouraged, undesirable ones discouraged, and risks reduced by subsidies, taxes, regulations, insurance, and communication of relevant information.

Clearly the need for wise public and private action is acute for the period ahead. The combination of complexity, important indirect effects, and economy-wide scope renders many strategic decision problems unusually resistant to obtaining good answers through judgmental analysis or fully decentralized decision-making. Quantitative models for forecasting and policy analysis are important for contributing to such decisions. Short-run prices may give poor guidance to long-term commitments in the private sector. Muddling through short-term incrementalism in the public sector may create serious, even irreversible, long-term external effects.

Quantitative modeling can reveal the future and improve our foresight and, where the prognostication is undesirable, simulate and test alternative strategies designed to produce coherent, mutually reinforcing public and private actions.

How prepared are we to respond to this need? To answer this question I will examine in greater depth the character of the energy adjustment process and the present state of the modeling art. Only if models can make accurate and relevant forecasts, or conditional forecasts of the outcomes of actions, will they be useful in addressing energy issues. The key words here are "relevant," "accurate," and "unconditional and conditional forecasts."

The Problems and an Ideal Modeling Approach

Examining the characteristics of an ideal model formulation and application illuminates the modeling approach to the energy problem. Ideal models may, however, substantially exceed present capabilities.

Dynamic Structure

It is useful to distinguish among five different adjustment processes to energy-induced changes occurring over successively longer time periods.

First, the response that an economy makes to an energy disturbance in a quarter or two is limited largely to changing: production rates, over-time, temporary layoffs and recalls, orders and shipping, and financial assets and liabilities.

Second, in a year an economy can respond by modifying production processes and machinery, by changing wages and prices (that usually are set with a view to long-term sales, recruiting, and turnover), by hiring or terminating employees, training, and by revising government regulations.

Third, in two years new plants can be built, households can relocate to new regions, and legislation can be passed.

Fourth, in four years new technology can be developed, designed into processes and equipment built, new sources of raw materials developed, and new infrastructure projects implemented by governments.

Fifth, in ten to twenty years plant and equipment will wear out or become technologically and economically obsolete, and raw material sites can be exhausted.

The response times above obviously differ markedly by industry, agency, etc. Behavioral, economic, and technological relations that are incorporated in model structures to reflect the above phenomena would be quite different for each of the five dynamic adjustment processes. Model parameters that are fixed in the short run become responsive variables in longer adjustment times. Since the social, economic, technological, and physical world actually embodies these increasingly fluid adjustment processes as time passes, any mathematical model should reflect the dynamic adjustments for the time spans during which it is intended to be valid. Otherwise it will not be able to make with any degree of accuracy the unconditional forecasts or conditional forecasts for policy analysis that are needed. While no model need encompass all of these dynamic structures, ideally a model accurately reflects the dynamic structure for the time needed for the energy question under study.

An ideal energy model starts with an econometric model estimated from historical data that reflected the current structure of the economy. With the passage of time, the models would gradually phase-in mathematical programming models. Such models introduce new technologies by constructing new plants built in response to increased energy prices.

Public and Private Sectors

Since the American economy, particularly the energy sector, involves a strong interaction between its private and public sectors, an ideal model incorporates the behavioral constraints of both. The corporate and household sectors of the model should reflect not only neoclassical profit and satisfaction motivation but also organizational behavior and information limitations. The public sector of the model should reflect the demand for services and revenue sources that constitute its economy and reflect bureaucratic organization, political processes, and constituency pressures, including the impact of wealth on political power.

These processes and structures influence public and private decisions, and ideally should be taken into account in the analysis of a socially desirable energy policy. Many policy analyses based on economic models have yielded recommendations that foundered in application because they ignored the political process and bureaucratic structure of government or the interactions between wealth and power.

Model Application

A model useful for energy analysis is likely to be so detailed and complex that it is easily confused with a general-purpose model good for all applications. That is likely to be a mistake since even our mythical ideal model still would be a grossly oversimplified caricature of reality. At best it will be adequate for only a specific range of issues.

The ideal model is designed for specific uses in order to be tractable and comprehensible to its user. No model ever tells the whole story so its use always involves the user's judgment. In use a model will simulate alternative courses of action whose outcomes are then subjectively evaluated. Alternatively when the user can enumerate objectives, the modeling system can concentrate on the most viable alternatives using maximization algorithms.

Research Centers for Modeling

Finally the ideal modeling effort is produced by a research center or other institution that, because of its own needs for support, security, and access to data, problems, and decision makers, is effective in building and applying models. To solve energy problems, the modeling institution needs access to both the public and private sectors, although close contact with one decision group may preclude similar intimacy with others.

State of the Modeling Art and Institutions

How accurately do available models reflect the dynamic structure of both public and private sectors on energy issues and how effectively are they used for making forecasts and policy analysis and communicating the results to the responsible decision makers? Because an in-depth answer would require much research, I will characterize briefly the present state of the art in broad terms. A few general observations will serve as background.

Overview

Quantitative analyses have long been made of specific energy sources in terms of reserves, new finds, utilization rate, and projected exhaustion dates. Usually these studies were made independently of other energy resources. Only in recent years has attention begun to focus jointly on energy of all forms. Cost considerations led to the expectation that when oil and gas ultimately are exhausted, usage would shift to coal, and finally to a virtually inexhaustible supply of atomic energy.

Prices in real terms were seen as steadily rising, but levels which we have already experienced were not foreseen. As the exhaustion of American oil resources began to appear, world oil models were formulated. The increasing U.S. reliance on Mideast oil was then foreseen, and the rhetoric of national policy set the objective of self sufficiency. While this was a plausible objective in military terms, it was never really squared with the facts that using Mideast oil was cheaper and left available larger reserves of domestic oil for military security.

About this time, large, complex modeling efforts were launched by the federal government that attempted to span all fuels, industries, and regions. These ambitious but loosely coordinated efforts involved many groups and laid the foundations for most current energy models. Evidently they have yet to influence national energy policy importantly, but nevertheless, much has been accomplished in making energy modeling a significant tool of analysis.

Modeling Dynamic Structures

Very large changes in energy prices induce correspondingly large changes in technology and in the structure of the economy. While all five different responses involved in these changes are not likely to be needed, most energy models give inadequate attention to structural change.

Often econometric, input-output, and industrial dynamics models are fitted to historical data using statistical methods that implicitly assume that the estimated parameters are constants. Alternatively, linear programming models sometimes are based on engineering estimates of processes that are potentially efficient. Optimization algorithms then invoke process changes in response to conditions prevailing in the future. These two classes of models appear to be either backward looking or forward looking, but do not make the transition from the former to the latter.

To be sure, some models try to span the past and future by distinguishing between short-term and long-term demand elasticities, by distinguishing between the determinants of current production and the investments in capital stock, and by making input-output parameters vary with changes in relative prices. All of these are relevant but fragmentary treatments of the important structural changes that rising energy prices are likely to induce in the mix of technologies and industries and in the location of production and population.

Forecasting energy changes and analyzing policy options usually will require long time spans that allow ample opportunity for the present economic structure to evolve into quite different forms. The present state of the art is quite limited in its theoretical treatment of the dynamic adjustment processes that change structure, and in the data bases for studying them. In recent years statistical estimators have been developed for estimating drifting parameters, but have yet to receive wide use (2). These estimation methods seem particularly appropriate for forecasting the diffuse impacts of energy-induced structural changes.

Most energy-oriented models today are primarily neoclassical economic models that pay little attention to technological innovation. Yet the dynamic processes of invention, development, diffusion, and investment will significantly influence adjustments to energy shortages.

Behaviorally oriented models of firms and households are needed that take account of information generation, communication, and processing. This point applies to businesses in energy production as well as to firms and households in consuming energy. The static equilibrium analysis that often allows economists to ignore the behavior of people will be inadequate in a dynamic world of fast-changing energy structure. The vacillation between big car and small car purchases in response to congressional debates about the reality of the gasoline shortage is a case in point; American automobile manufacturers were seriously misled in estimating consumer intentions.

Modeling Public-Private Interactions

Government-sponsored modeling examines the interface with the private sector in terms of private responses to governmental instruments such as taxes, subsidies, etc. Seldom is the public process modeled so that the decision makers receive help in judging political and administrative issues. Since the political actors in the public policy process are jealous of their prerogatives, the role of models as an aid in the decision making process must be emphasized. Models should not be seen as a substitute for decision makers, who must face the remaining uncertainties and make the final decision.

Joint analysis and modeling of the public-private sectors is important, because the energy problem should not be treated as simply a narrow economic issue. Who will be hurt and helped and where they live are indispensible aspects of political feasibility.

Conceptual models (3) have been developed for delineating the respective roles of government and business in a mixed economy regulated by both market and administrative processes, but they have yet to be effectively used in policymaking processes. Current policy analyses reflect this limitation and suffer from it.

In a problem area such as energy, modeling should, to the extent possible, be governed by the problem which is jointly economic, political, and technological, rather than following the traditional disciplinary divisions of the university. By this standard, today's models are deficient, particularly in neglecting the behavior of the public sector.

Model Applications

Forecasting applications of models are relatively straightforward although the theory of lag structures is thin and simple, and statistical models often do almost as well as complex structural models. Political and technological imponderables and drifting structures limit severely the forecast accuracy of current models, so it is impossible to rely on them very far into the future.

In some respects policy analyses are less demanding than forecasting since conditional forecasts can be made for incremental changes in policy actions. Forecasting this incremental response may well be insensitive to an accurate estimate of structure. Current models are often used in this way with considerable confidence in the results.

When the decision maker supplies judgments of social benefits, simulation studies of alternative policies can be made using currently available models. However, if more objective measurements of benefits are desired, we encounter a badly neglected area. Research on the estimation of utility or social welfare functions is still in a primitive state. Some means of correcting this deficiency is essential if the power of the computer is to be used by the decision maker in searching policy alternatives.

Interface between the modeler and the decision maker usually is a troublesome barrier to successful applications because of the differences in their functions, their knowledge, and their time scales (4). The modeler would require a two-year head start to achieve the instantaneous response that the decision maker wants. Where problems are well structured, the decision maker in many cases has been able to work effectively with the operations research/ management scientist in specifying objectives and constraints and in translating the results of quantitative analysis into action.

Unfortunately most energy policy issues are not well structured but instead are complex and amorphous. Modeling does not yet offer a methodology for achieving strategic analysis that will enable decision makers to deal with problems in orderly, comprehensible stages in which models and judgment are balanced and combined. Decision trees, dynamic programming, and hierarchical structures are all relevant components of an evolving but still inadequate methodology.

When such complex problems occur in the public decision arena, modelers are even less useful than in the private sector where the decision making responsibility is centralized. Public decision procedures typically utilize adversary structures that lead to independent single-interest analysis and bargaining positions that are resolved by power and compromise (5). Integrative solutions that simultaneously take account of all objectives are seldom even formulated. Quantitative modeling and decision analysis can make an important contribution in this area, but has yet to do so. Such capabilities are badly needed for such problem areas as energy.

Research Centers for Modeling

Effective models will require applications in research centers that engage in multidisciplinary, large-scale, long-term research. These programs encompass wide-ranging work: basic and applied, quantitative and institutional, and behavioral and technological. Such research must be supported by extensive data sets and powerful, flexible computing capabilities. These capabilities are not provided by the discipline-determined structure of independent university departments.

Applied research centers inside and outside universities have aspired to these capabilities in the energy area and have attained some success. But the support for these efforts usually has been on a small-scale, project-by-project basis that is not conducive to sound institution building. The result has been too many underfunded, insecure institutions pursuing fragmented efforts and devoting inordinate energies to mere survival (6).

The erratic pattern of contracting out model projects prevents building centers for modeling research that have interfaces with government, industry, universities, and the public. Such centers should provide communication, credibility, and a balance of independence and responsiveness.

Most of the available resources have consequently been devoted to building models, and until recently little attention was given to validation and adequate documentation of models, computer programs, and data sets. Communication, transfer, utilization, and credibility have suffered as a result.

Conclusions

This paper provides an overview of our energy problems, the needs for quantitative modeling, the characteristics of an ideal approach to evaluation, and an assessment of the present state of the art of energy modeling. My objective was a broad perspective of where we are and where we might go. Gross deficiences were found in several areas:

(1) Most models have not adequately dealt with energy price-induced changes in the form of: the generation and diffusion of new technology, its gradual embodiment through investment, industrial growth and decay, and the migration of population. Adequate modeling of long-run adjustment dynamics will require integrated models that incorporate existing structure in econometric relations. As the time horizon is extended using technology-based programming models, the models introduce new activities when they become economically efficient.

(2) For models to be most helpful in formulating and implementing energy policy, they should reflect the costs and benefits that will constrain the political decision process, and the administrative capabilities and costs of operating various types of programs. Incorporating these in the modeling structure will allow such considerations to enter the policy analysis at an early stage; efforts can be concentrated on supplying options and relevant information to decision makers.

(3) Because energy issues involve the active participation of both public and private actors there is an urgent need for clarifying their respective responsibilities. This is important because both the economic and technological uncertainties and the political conflicts

of interests that already are involved make energy program decisions quite difficult to achieve even apart from resolving the division of public and private responsibilities. Much more empirical and theoretical research is needed on determining objectives and on decision strategies for coping with complex policy problems.

(4) Adequate and continued support is essential for developing the sizable research institutions required for building, testing, and maintaining large models. The urgent challenge of energy issues underlines this long-recognized need.

American researchers have accomplished a great deal in developing modeling capabilities for the energy problem--enough to demonstrate its potential. We still have a long way to go in achieving the knowledge, methodology, and institutions that are required for full effectiveness.

Footnotes

1. Surveys and analyses dealing with specific energy models and issues are available in Olson (1980), Energy Modeling Forum (1977), Charles River Associates (1979), and Manne (1979).

2. See Rosenberg (1973) for a survey.

3. See Musgrave (1959).

4. See Henderson (1980) for an approach to the interface problems.

5. See Lindblom (1959) for an interpretation of current public decision making in the context of complexity.

6. For elaboration see Greenberger (1976).

References

1. Charles River Associates, Review of Large Scale Energy Models, EPRI Report No. EA968, Palo Alto, January 1979.

2. Energy Modeling Forum, Energy and the Economy, Stanford: Institute for Energy Studies, Stanford University, September 1977.

3. Greenberger, M., Models in the Policy Process. New York: Russell Sage Foundation, 1976.

4. Henderson, J.R., R.R. McDaniel, and G.R. Wagner, "The Implementation of Operations Research/Management Science

Modeling Techniques," IEEE Transactions on Engineering Management, 1980. EM-27, 1, 12-18.

5. Lindblom, C.E. "The Science of Muddling Through," Public Administration Review, Vol. 2, No. 4, 1973. p. 381-398.

6. Manne, A.S., R.G. Richels, and J.P. Weyant. "Energy Policy Modeling: A Survey," Operations Research, Vol. 27, No. 1, 1979.

7. Musgrave, R.A., The Theory of Public Finance, A Study in Public Economy. New York: McGraw Hill, 1959.

8. Olson, J.A., T.R. Plaut, and C.C. Holt, Survey of Regional and Energy Models, Final Report prepared for the Texas Energy and Natural Resources Advisory Council, Bureau of Business Research, The University of Texas at Austin, April 1980.

9. Rosenberg, B., "A Survey of Stochastic Parameter Regression." Annals of Economic and Social Measurement, Vol. 2, No. 4, 1973. p. 381-390.

Index